Introduction to Building Procurement

To my wife, Sheila, for her constant support, love and encouragement, Brian.

To friends old and new, Graham.

Introduction to Building Procurement

Brian Greenhalgh and Graham Squires

 Spon Press
an imprint of Taylor & Francis
LONDON AND NEW YORK

This edition published 2011 by Spon Press
2 Park Square, Milton Park, Abingdon, Oxon OX14 4RN

Simultaneously published in the USA and Canada by Spon Press
270 Madison Avenue, New York, NY 10016, USA

Spon Press is an imprint of the Taylor & Francis Group, an informa business

Typeset in Sabon by Bookcraft Ltd, Stroud, Gloucestershire
Printed and bound in Great Britain by TJ International Ltd, Padstow, Cornwall

British Library Cataloguing in Publication Data
A catalogue record for this book is available from the British Library

Library of Congress Cataloging in Publication Data
Greenhalgh, Brian.
Introduction to building procurement / Brian Greenhalgh and Graham Squires.
   p. cm.
   1. Building materials--Purchasing. 2. Building--Estimates.
   3. Construction industry. 4. Building materials industry. I. Squires, Graham.
   II. Title.
   TH435.G66 2011   690.068--dc22
   2010029251

ISBN13: 978-0-415-48215-8 (hbk)
ISBN13: 978-0-415-48216-5 (pbk)
ISBN13: 978-0-203-88506-2 (ebk)

100623025 1

Contents

Figures

Tables

Foreword

As this new text makes clear in its opening paragraphs, the construction industry is very important to an economy, and the data and facts referring to the UK could equally apply to many high-income economies. Indeed in most European countries and American states construction activities now account for approximately 5 to 10 per cent of a nation's GDP and are typically responsible for the employment of millions of site labourers and professionals. Obviously these figures are sensitive to the general pattern of economic activity. For example, the £30 billion cuts in UK public spending announced in the first coalition budget in June 2010 (the week this foreword was written) could seriously jeopardise the volume of construction output seen in 2011 and beyond, especially as the government procurement arm, the Office of Government Commerce, was merged into the Cabinet Office as part of a cost-cutting efficiency drive in the very same budget!

The central point made by this historical commentary on the current economic climate is that to make sense of it, at any point in time, requires economic literacy, and part of this inevitably means that you need to appreciate the nature of the economic transactions that underpin each and every new order for a constructed product. As this *Introduction to Building Procurement* makes transparent, the world of construction is complicated by the initial negotiation between client and contractors, as no two jobs are quite the same and as a consequence the quality of a project, the time it takes and the cost involved are often unknown variables at the start of the procurement process. As the authors perceptively state at the very beginning of their text: 'From the client's perspective the investment is unpredictable in terms of delivery, budget and the standard of quality.'

From my position as a construction economist the two subject areas appear inseparable, but as you can see there is too much to each specialism to fit comfortably into one text. As a comprehensive work on procurement, however, you will find it difficult to better what is offered in this book. The breadth of experience that it draws on is certainly unique, and together the authors have provided a detailed and clear exposition of construction procurement. The text began its life when Jack Masterman was studying for his PhD in the early 1990s; Brian Greenhalgh subsequently recognised from his significant experience in the industry and academia that this needed to be revised and extended to take account of the various developments of the twenty-first century. Graham Squires, at the other end of his career, contributes an energy and sense of purpose to help bring the long period of gestation to publication. Between them they have produced an interesting and worthwhile contribution that I can *happily* say you will understand. The term 'happy' may appear subjective and unusual in academic commentary, but from my perspective this text should enable its readers to begin to appreciate the complexities and oddities of construction markets and hence put them in a stronger position to become more rounded property and construction professionals.

Danny Myers

1 Introduction and nature of the construction industry

The construction industry and projects forming the built environment involve many activities. According to Section 105 of the Housing Grants, Construction and Regeneration Act (1996), a construction project can refer to any building activity that includes alteration, repair, erection, demolition, maintenance, painting, land clearing, earth moving, grading, excavating, trenching, digging, boring, drilling, blasting, concreting and installation of machinery. The construction sector is very important to both the UK economy and other Western economies, as in most years it accounts for approximately 8–10 per cent of gross domestic product (GDP). In the UK more specifically the construction sector employs around 1.5 million people. The falling value of building procurement has been mirrored in the slow down of building activity since the most recent recession. In the first quarter of 2009 only £258 million worth of deals were recorded in the building sector compared to £7.2 billion during the first quarter of 2007. Future upturns in the economy will no doubt see a return to growth and procurement in the sector, although a return to the levels of output experienced prior to recession is uncertain. As a result, the significance for building procurement during negative or slow economic growth will be its drive for efficiency, effectiveness, and value for money.

The construction sector is to a large extent investment-led, which means that when a client procures a building, they are buying an 'asset' which has the ability to generate funds into the future. Approximately 50 per cent of building work is generated from private clients and 50 per cent from central and local government, so it is inevitable that the industry's output will fluctuate in accordance with the economic and political cycles. It is generally acknowledged that cycles in building activity show far greater amplitude than cycles in other business activity over the same period of time. For instance, the industry took a distinct upturn in the late 1990s in response to government infrastructure investment programmes. Government-developed investments were those such as the private finance initiatives (PFIs) in health, education and transport. These PFI programmes are discussed in more detail in Chapter 11.

The building construction industry covers a wide range of business activities that are brought together by a common interest in the development of land and real estate. The sector comprises a variety of interests that include clients, designers, suppliers and contractors. Clients can be industrial firms and commercial property developers who determine what should be built and where. Others might be government departments who require various forms of infrastructure to support the residents and communities that they represent. Some of the key building clients are the Environment Agency, British Waterways, British Energy, Railtrack, the Highways Agency, port authorities, water companies, major retailers and house builders. A more detailed discussion of the nature of clients in the industry will be found in Chapter 2. Designers have an interest in the building sector by determining the detail of what should be built, including the size, shape and specification of the finished article. Suppliers provide materials and components via the processes of extraction and manufacture. However, it is the business activities of contractors in carrying out building operations that inevitably contribute most to the development of land. These widespread activities of clients, designers, suppliers and contractors demonstrate the fragmented nature of the construction industry. Within this context of fragmentation the complex nature of procurement in the construction industry will now be introduced.

1.1 The nature of procurement in the construction industry

A by-product of this level of complexity in building procurement is the large number of business transactions that need to be set up to support just one project. The transactions themselves are not the distinguishing feature, as virtually all economic transactions rely on the purchase of numerous goods and services. In everyday parlance this is simply referred to as 'buying'. In building circles, however, the purchase of goods and services is commonly referred to as *procurement*, and the design and operation of efficient mechanisms for procuring goods and services has become an important area of study. Indeed, there are enough systems of procurement used in building to fill a textbook.

Buying standardised goods and services is a relatively straightforward exercise, as a standard bidding or tendering process can be normally relied upon to produce an efficient outcome in a competitive market. However, the procurement of building is frequently more complicated. This complexity is often due to the unknown variables at the start of the project such as the quality of the job, the time it will take to complete, and the cost to the client. From the client's perspective the investment is unpredictable in terms of delivery, budget and the standard of quality. For example, a high-quality polished timber or steel façade may entail a cost that is unknown when the work commences and may require a special skill that is not fully understood by the client. Further unpredictability may occur if the work takes a considerable amount of time to prepare and if the quality of the end product is not easy to verify or quantify.

This book addresses how these problems of specification, price, project duration and value for money have been tackled by clients and suppliers of building goods and services in the public and private sectors. Those who can learn from textbooks like this will also help make the sector more professional. The building industry is characterised by a number of irregularities, such as collusion, less than rigorous tendering and poor workmanship. For instance, there is considerable potential for those involved in the procurement process to favour certain suppliers, so by focusing attention on the problems and irregularities of the sector, a progressive shift in the 'nature' of the profession will improve both the level of professional competence and actual work carried out in it.

1.2 The structure of the building industry

1.2.1 The building industry in the economy

Building output in the UK increased each year between 1995 and 2010, benefiting greatly from the expansion in the UK economy in that period. Activity in building was especially robust between 2000 and 2004, with annual expansion averaging over 4 per cent. However, output fell in 2005, mainly because of a depressed housing market and reduced spending in the public sector as a result of financial constraints. This depressed trend continued in 2010 and was dampened further by cuts in the public sector.

Historically, building (excluding new housing) has grown in line with GDP but with much sharper peaks and troughs – particularly in the late 1980s and early 1990s. However, since 1995 this relationship appears to have changed, and the industry has become slightly counter-cyclical. This reversal may be due to the emergence of longer-term initiatives such as facilities management (FM) and PFI that have had the effect of smoothing out the cycles. Greater London, the south-east of England and the east of England have enjoyed growth in output since the mid-1990s. However, building output in Greater London fell in real terms by 5 per cent between 2003 and 2005. This was partly due to a reduction in major infrastructure projects and has since been reversed following the award of the 2012 Olympics to London.

1.2.2 Absence of strong overseas competition

The UK's building industry is geographically offshore from mainland Europe, and therefore does not experience the full effects of cross-border competition on its domestic market. The historical absence of strong overseas competition in the domestic market (outside building products) has resulted in a culture that does not perceive the imperative for change. This is in contrast to experience gained by other, more internationally competitive UK industries in the post-war period. The gap between the best, which is world class, and the average is thought by some to be too great (Department of Trade and Industry 1998). This domestic focus in the nature of the UK building industry is apparent in the many UK controlling bodies and societies that shape its development.

1.2.3 Key shapers

In the UK, the Communities and Local Government Department (CLG) heavily influences planning and building policy. As far as building procurement is concerned, CLG affects activities such as the planning system, decisions on planning applications and appeals, and building regulations. In England, local authorities are appearing to have

Table 1.1 International building output growth (percentage on year ago) 2002–8

	2002	2003	2004	2005	2006	2007	2008
UK	3.7	3.7	6.0	–1.2	2.5	3.6	4.7
Western Europe	–0.2	0.7	2.1	1.1	1.3	1.4	1.6
Eastern Europe	1.7	2.8	6.1	5.4	7.3	7.5	8.2
US	5.0	11.0	8.0	3.0	3.0	5.0	5.0
China*	8.8	12.1	8.1	10.5	8.4	7.0	8.2

Source: OGC Building Demand/Capacity Study, June 2006

Note:
* Data for China are value added from national accounts, other series are gross output

more powers in providing regulatory control in planning and building. Scotland, Wales and Northern Ireland also have their own systems, regulated by their Parliaments or Assemblies. Bodies such as the British Property Federation (BPF) represent the owners and investors in commercial and residential property. Property assets held by BFP members are currently worth over £70 billion, so it plays a leading role in representing the interests of the clients of the building industry, and liaises closely with the government via its building committee.

The various professional institutions play a key role in both representing their members and influencing policy. In the construction industry, the main bodies are: the Royal Institute of British Architects (RIBA), the Royal Institution of Chartered Surveyors (RICS), the Chartered Institute of Building (CIOB), the Institution of Civil Engineers (ICE) and the Chartered Institute of Building Services Engineers (CIBSE). These are the most senior institutions as they have 'Royal Charter' status and therefore have access to government decision-makers, although there are many other institutes representing more specialist areas of the industry.

Overarching these specialist bodies is the Construction Industry Council (CIC), the representative forum for many of the professional bodies, research organisations and specialist business associations in the construction industry. It is designed to provide a single voice for professionals in all sectors of the built environment through a collective membership of over 500,000 individual professionals and 25,000 firms of construction consultants. In addition, the CIC represents the views of the higher level of the industry (professional, managerial and technical) in an organisation called Construction Skills, which is a Sector Skills Council for construction, and a partnership between CIC and the Construction Industry Training Board (CITB).

1.3 Stages in a building project: from conception to delivery

A building project can take many forms, as each individual project is unique. Each building site is inherently different, with its own complex characteristics such as ground conditions, surface topography, weather, transportation, material supply, utilities and services, local subcontractors, labour conditions and available technologies. However, there is a logical standard order of stages that need to be considered for each individual project undertaken. The tables below provide important detail and an activity checklist of the typical stages and action stages carried out in a building project (Tables 1.2 and 1.3). Building project stages include pre-design, design, bidding/negotiating, building, operation, delay and abandonment. Practical actions will need to be taken during each of these stages (see Table 1.4). In progressing from project planning to project completion each stage requires input from a variety of contributors and stakeholders. Contributors include the project participants, such as client, designer, design team, contractor and supply chain. Stakeholders include financial organisations, government agencies, insurance companies and other non-direct organisations that still have an interest in the project. Note that further discussions of all the roles carried out in a project can be found in Chapter 2 and Chapter 4.

These stages in Tables 1.2 to 1.4 offer parallels to the *Outline Plan of Work* drawn up by the Royal Institute of British Architects (RIBA 2007). This RIBA *Plan of Work* is discussed in more detail in Chapter 6, as it sets out the process to be followed for projects to be procured mainly using the separated procurement systems carried out by traditional means. To provide a brief insight and demonstrate the overlap, the RIBA *Plan of Work* identifies twelve stages, which include: inception; feasibility; outline proposals; scheme design; detail design; production information; bills of quantities; tender action; project planning; operations on site; completion; and feedback. Connections to the model below (Table 1.2) would draw on all stages that include pre-design (drawing on the RIBA stage of 'scheme design'), design (drawing on the RIBA stage of 'detail design'), bidding/negotiating (drawing on the RIBA stages of 'outline proposals', 'production information', 'bills of quantities' and 'tender action'), building and operations (drawing on the RIBA stages of 'operations on site', 'completion' and 'feedback'). It is interesting to note that delays and possible abandonment in project stages are not considered in the RIBA model. However, this issue of delay in a project is covered in more detail in the course of the discussion of the RIBA work plan in Chapter 6.

Table 1.2 Stages in a building project

Project stage	Stage detail
Pre-design	• Has not yet advanced to design nor has a design team been selected. • The project could be awaiting financing, land acquisition, review agency approvals, etc. • Typically the only contact listed on a project at this stage will be the owner or the owner's representative.
Design	• Used on reports that list an architect or primary design factor. • A design team has been selected at this point and design is under way. • Design stage reports are usually issued prior to bidding, negotiating and/or start reports, although a project will be assigned a design stage along with bidding, negotiating, start or building if design is still in progress while the project is bidding or under building.
Bidding/negotiating	• Indicates that the owner is accepting bids (prices) from general contractors, subcontractors, suppliers and manufacturers. • Plans may or may not be complete while bidding is in process. • When this is the case, you may see a dual project stage selection such as bidding and planning/final planning as some phases of the project may be bidding while other phases are still under design.
Building	• Work is either under way or is scheduled to begin within 60 days. • The name of the general contractor, builder, contractor, or the fact that the owner or some other factor listed on the report will subcontract the project will also be listed on building reports. • Start of work is defined as the start of site work to accommodate the building foundation or later activity. • Site work, not site clearance or demolition, is considered the start of the project as long as it is expected to begin within 60 days.
Operation	• Notice of completion is a key feature of this stage.
Delayed	• Progress on the project has been delayed and has come to a standstill at this point in time. • This may occur at any time during a project's life cycle but usually will be seen during the design stage.
Abandoned	• Assigned to a project that has been verified through the owner as no longer viable and is not going to move ahead. • The project has effectively been killed.

Source: McGraw-Hill Construction 2010.

Table 1.3 Stages and action stages in a building project

Project stage	Action stages
Pre-design	**Request for proposals** This action stage is used to indicate that the owner is seeking proposals from a design firm (usually an architect or engineer). **Request for qualifications** This action stage is used to indicate that the owner is seeking qualifications from a design firm (usually an architect or engineer). **Pre-design** This action stage is used for all other pre-design information other than a request for proposals/qualifications from an architect or engineer.
Design	**Planning schematics** Design is still in the early stages, and the project is not expected to bid or start building for more than four months. **Design development** Design is well under way, and the project is not expected to bid or start building for more than four months. **Building documents** The project will be bid or advance to building within four months. This usually corresponds with advancement of the project design to building documents or working drawings. **Pre-qualification** A general contractor or building manager is being asked to submit qualifications and bidding/building is expected within four months. If the owner is seeking General Contractor (GC)/Contract Manager (CM) qualifications and the project will not advance to bid/building in four months or less, the project will have a dual action stage of bidding with the appropriate planning schematics or design development action stage, depending on the advancement of the plans. If no architect or engineer is involved in the project before the GC/CM pre-qualification occurs, the project will be issued with the dual action stage of bidding/pre-design.
Bidding/ negotiating	**Bidding** Bids are sought on an individual trade, material or piece of equipment, on a series of bid packages, or on an engineering project. **General contractor bidding** Owner is publicly seeking GC bids, open to all interested and qualified firms. **General contractor bidding invitation** Owner is privately seeking GC bids from a select list of invited firms. **Sub-bidding** A GC or CM is seeking bids on individual trades or trade packages. Sub-bidding will always be accompanied by another action signal indicating the status of the project. **Negotiating** Owner is negotiating contract with one or more GCs. Dodge defines negotiating as the owner dealing with a limited number of contractors, usually two or three, on an informal bidding basis. It is important to remember that while a project is being negotiated, the prime contractors involved in the negotiations still go through a very similar process to the one they go through if they are bidding the project. The negotiating contractors must still get bids from subcontractors, suppliers, dealers and distributors. Negotiating a GC/CM contract may occur during or after the design stage.

Project stage	Action stages
Bidding/ negotiating *continued*	**Bid results** The bid result project stage is used when reporting the results of a bid opening, rejection of bids, bids received unopened, bids returned unopened, and bids 'in' status. A bid result project stage used in conjunction with a start project stage usually indicates that there will be no further updates reported on the project.
Building	**Start** The first report issued to indicate that work is under way or is scheduled to begin within 60 days. **Subcontract award** Indicates that subcontractors who have been awarded a contract are listed on the report. This stage is always used in conjunction with another action stage, indicating the status of the project. **Permit** Assigned to projects that are being reported in start as a result of information gathered from a building permit. **Building** All subsequent reports that indicate that work is under way or is scheduled to begin within 60 days after the start report.
Operation	**Notice of completion (NOC)** • Is assigned to projects that have reached the point of active operation or leasing. • Does not routinely follow projects to this stage and you only see it listed if the project had not previously been reported in the start/building stages. **Leasing** See Notice of completion. **Service bidding** See Notice of completion. **Retrofit** See Notice of completion.
Delayed	See Notice of completion.
Abandoned	See Notice of completion.

Source: McGraw-Hill Construction 2010.

1.4 Value for money in competitive markets (time, cost and quality relationship)

The level of trade-off between time, cost and quality will determine a project's value for money in a competitive market. Government agencies have recently started using new types of contracting methods which have placed an increasing pressure on decision-makers in the building industry to search for an optimal resource utilisation plan that minimises building cost and time while maximising its quality (Cristobal 2009). Figure 1.1 demonstrates how movement outwards from the centre of the triangle to one of the points (e.g. time) will mean that a trade-off in other variables will have to be forgone (e.g. cost and quality).

Table 1.4 Actions required during the stages of a building project

Project stage	Action to take
Pre-design	• Building product manufacturers Contact owner to pre-sell or negotiate agreements on products or services. • Architects/civil engineers Contact owner to seek design opportunities. • General contractors or design/build firms Contact owner to persuade them to use your project delivery system. • Banks, financial institutions, and mortgage companies Contact owner to offer financial assistance. • Real estate companies Contact owner to help with site or tenant selection.
Design	• Contact the design source to review product applications, obtain product approvals, or influence the specifications. • Provide the project manager with pertinent product literature. • Contact the owner to pre-sell or influence product selection • Seek follow-up time frames. • Determine if the project is right for them and pass up marginal jobs. • Use personal calls or direct mail to attempt to negotiate a contract. • Obtain product approval and specification. • Get prequalified for a bid list. • Adjust schedule for bidding or negotiations.
Bidding/ negotiating	• Contact the design source to review product applications, obtain product approvals, or influence the specifications. • Provide the project manager with pertinent product literature. • Contact the owner to pre-sell or influence product selection • Seek follow-up time frames. • Determine if the project is right for them and pass up marginal jobs. • Use personal calls or direct mail to attempt to negotiate a contract. • Obtain product approval and specification. • Get pre-qualified for a bid list. • Adjust schedule for bidding or negotiations. • Contact the negotiating general contractor to submit a price on trade or material. • Obtain subcontractor names • Contact low bidder and attempt to get a firm commitment • Request names of favoured subcontractors
Building	• Seek successful subcontractors. • Subs will make final try with revised bids or offers. • Close out activity files. • Prepare list of active and successful GCs to tailor promotional efforts. • File by competitor (knowledge of workload will help in future bidding). • Market analysis statistics. • Award sub-trades. • Begin work.
Operation	• Obtain Notice of completion
Delayed	N/A
Abandoned	N/A

Source: McGraw-Hill Construction 2010.

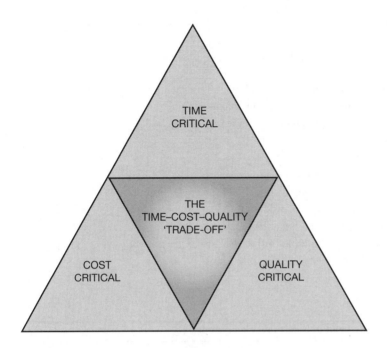

1.1 Time, cost and quality relationship triangle

Several subcategories of time, cost and quality can also be considered in this relationship. For instance, time may involve not just the actual time or duration in construction, but elements of predicted time in constructing if there are delays or contingencies to add. It may be that there is a chance that a client wants to vary their requirements during the course of the project. These predicted as well as actual elements are also considered for costs, as potential changes in building during the project will no doubt increase the overall final cost of the project. As for subcategories of quality, elements of defect will be important, and other issues associated with quality will be different at differing periods of the project timeline. For example, quality standards can be measured up to a certain point, such as when a property is available for use or at the point when the contract states that defects should have been rectified. These elements of quality that encroach on aspects of time demonstrate the complex nature of measuring individual variables as separate entities, although consideration of their interrelation is critical to ensuring that value for money is obtained as part of the procurement process.

1.5 Supply and demand

Within a modern market system the appropriate factors of production (land, labour, capital and entrepreneurship) are employed to create goods and services. Within this market system, the laws of supply and demand affect the price and the quantity of goods and services. In doing so, laws of supply and demand bring together both producers and consumers. In unfettered or perfect markets, these goods and services

Table 1.5 The time, cost and quality relationship

Time	Time for building
	1 Time predictability – design
	2 Time predictability – building
	3 Time predictability – design and building
	4 Time predictability – building (client change orders)
	5 Time predictability – building (project leader change orders)
	6 Time to rectify defects
Cost	Cost for building
	1 Cost predictability – design
	2 Cost predictability – building
	3 Cost predictability – design and building
	4 Cost predictability – building (client change orders)
	5 Cost predictability – building (project leader change orders)
	6 Cost of rectifying defects
	7 Cost in use
Quality	Defects
	1 Quality issues at availability for use
	2 Quality issues at end of defects rectification period

Note: see also Table 3.7 in Chapter 3.

are produced (the supply) in equal accordance (in equilibrium) to the needs and wants of the consumers (the demand). To interpret market signals perceptively in the building industry it is necessary to consider the market, or price mechanism, in more detail. First, we need to look at supply, and more specifically the supply of goods and services in building projects. This will improve the way that resources are procured to meet differing supply requirements in the industry.

1.5.1 Supply of building

The total supply of building output in the UK is broken down into specific activities according to monetary value. House builders supply approximately 14 per cent of the annual output. Contractors supplying industrial and commercial buildings account for approximately 24 per cent of the total output in value terms. The large building and civil engineering firms undertaking complex infrastructure projects such as motorways, power stations and other public-sector activity account for a further 16 per cent of the industry's supply. This leaves the large number of small general builders dealing with repair and maintenance contracts to supply the lion's share, 46 per cent, of the activity (Building Statistics Annual 2009).

There are approximately 164,000 firms in the UK supplying building-based activity; therefore the industry is not one simple aggregated market. There can be little competition in the supply of products between the local builder undertaking repair and maintenance in a small town and a large national civil engineering firm. To avoid oversimplification we often need to envisage the building industry as having several different market sectors providing different types of supply. In this chapter the market for building projects (as an aggregate) has been used for illustrative purposes. This then sets the theoretical basis for application in different market sectors.

Theoretically, then, the supply of building goods and services is that which is made available by producers at a certain price and quantity. Examples of building supply include land as a site of building and labour such as tradesmen. Moreover, building supply could be the materials required for the actual building project, and the capital available could be in the form of machinery and finance. With supply it is always useful to distinguish between cost and price. Normally, the producer seeks to make a profit, where the cost of the good is less than the selling price. It is quite usual in building for the cost of a project to be estimated by the builder and a mark-up for profits and overheads added before arriving at a price for the job. The contractor's mark-up is the difference between price and cost. The process is complicated further by the fact that for most building work a price needs to be stated before the activity commences. The most usual form of price determination in the building industry is through some form of competitive tendering. This generally encourages the lowest bid to win the work. Chapter 5 provides a fuller discussion of tendering and payment procedures, while the rest of the book discusses the different methods procurement, of which competitive tendering is only one.

Economic convention states that the supply curve slopes upwards from left to right, demonstrating that as the price rises the quantity supplied also rises (Figure 1.2). Conversely, as price falls, the quantity supplied falls. As such, the basic law of supply can be stated formally as: the higher the price, the greater the quantity offered for sale; the lower the price, the smaller the quantity offered for sale, all other things being held constant. The law of supply, therefore, tells us that the quantity supplied of a product is positively (directly) related to that product's price, other things being equal. In using a building procurement example, the number of potential contractors interested in bidding to supply a project will increase as the profit margin the client is prepared to accept (and therefore price offered) rises.

1.5.2 Non-price determinants shifting the supply curve

Non-price determinants of supply are important and are demonstrated as a shift in the supply curve. This represents a change in both supply price and supply quantity. Non-price determinants include technology, government policy, supply-chain management and expectations. We shall now broadly consider four of these non-price determinants in turn.

First, if technology improves, the costs of inputs would become cheaper, resulting in a greater quantity of the goods or services to be supplied. For instance, technologically advanced prefabricated building production would result in a housing supply that could increase given the same market price, other things being equal. This would be demonstrated diagrammatically as a shift in the supply curve to the right (Figure 1.3). Second, government actions such as taxation and subsidies may effect a shift in the supply price and supply quantity. For example, the relatively recent landfill tax has increased building costs and reduced supply at each price. A subsidy would do the opposite by increasing supply at each price, since every producer would be 'paid' a proportion of the cost of each unit produced by the government. Direct legislation by government can also restrict or release supply in the market, and could be in the form of statutory regulations on building, planning or health and safety.

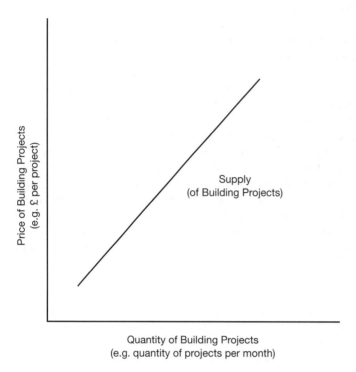

1.2 The supply of building projects

Third, supply-chain management is another non-price determinant of building supply. Most building activity normally involves integrating and managing many activities to reach the final product. Larger firms and conglomerates can for instance subcontract or diversify into other businesses to extend their range of operations. For example, a building firm may choose to merge with or take over its material supplier to guarantee that it meets completion targets on time. Such mergers may also reduce supply-chain costs and eliminate many of the associated transaction costs. The final key non-price determinant is expectations, and more specifically how expectations about future prices (or prospects of the economy) can also affect a producer's current willingness to supply. For example, builders or developers may withhold from the market part of their recently built stock if they anticipate higher prices in the future. This happened in the 1960s with the Centre Point building on Oxford Street in central London and in the 1990s with Exchange Flags in Liverpool. Most other cities can also offer similar examples.

By following along the horizontal axis, we can see that this rightward movement represents an increase in the quantity supplied at each and every price. For example, at price P, the quantity supplied increases from Q to Q_1. Note that if, on the other hand, the costs of production rise, the quantity supplied would decrease at each and every price and the related supply curve would shift to the left.

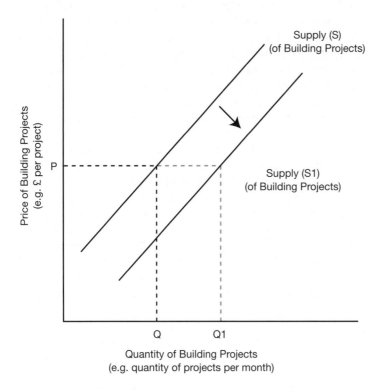

1.3 An increase in supply of building projects (a shift of the supply curve)

1.5.3 Inelastic supply in building

As well as determinants that shift both price and quantity, economists are often interested in the degree to which supply (or demand) responds to changes in price. The measurement of price responsiveness is termed price elasticity. Price elasticity is defined as a measurement of the degree of responsiveness of demand or supply to a change in price. For example, if a 10 per cent increase in price leads to a 1 per cent increase in the quantity supplied the price elasticity of supply is 0.1. That is a very small response. There are in effect three types of measure that economists use as a reference point to discuss price elasticity.

a Price-inelastic supply occurs when the numerical coefficient of the price elasticity of supply calculation is less than 1, supply is said to be 'inelastic'. The introductory example in which a 10 per cent increase in price led to a very small response in supply suggests a price-inelastic response: the measured coefficient was 0.1. Normally, price-inelastic goods are those that need to be bought irrespective of the price; an example in construction would probably be cement. All concrete and mortar needs cement and all buildings use concrete and mortar to some degree.

b   Price-elastic supply occurs when the numerical value of the price elasticity of supply calculation is greater than 1, supply is said to be 'elastic'. For example a small change in price elicits a large response in supply. This would be unusual occurrence in the markets for building or property but not impossible.
c   Unit-elastic supply is the most hypothetical case, as it describes a situation in which a percentage change in price leads to an identical percentage change in supply. This will always produce a coefficient value of 1. Again, this would be an unusual occurrence in the markets for building or property, but not impossible.

In the short run, the movement in price of any building goods or services tends not to proportionately affect the quantity supplied. The chief reasons for this are that a building project takes time to build and that building projects cannot quickly (if at all) be located to another area of market need. For building it is common to talk about short-run and long-run supply. The short run is defined as the time period during which full adjustment to price has not yet occurred. The long run is the time period during which firms have been able to adjust fully to the change in price. As an example, in the short run, rental values and house prices are demand-determined because adjustments cannot quickly be made to the supply of property. The markets for building in the short term are therefore price-inelastic in supply. It is this inelastic supply relative to demand that causes property markets to be unstable and characterised by fluctuating prices. In the extreme short run, the supply of buildings or infrastructure is fixed and the supply curve would be a vertical straight line. The example in Figure 1.4 shows how producers supply 200,000 building projects no matter what the price. For any percentage change in price, the quantity supplied remains constant (at 200,000 building projects) or inelastic. As mentioned, this feature of supply inelasticity is particularly notable within property markets as land is characterised by being perfectly inelastic. This is drawn from the notion that land is largely immovable. Existing areas of land can change use and character, but only in the long run as land is developed and property is built.

1.5.4 Demand for building

The demand for building and construction goods and services relates to the consumption of those products at a certain price and quantity. As Figure 1.5 shows, the demand curve for most goods and services slopes downward from left to right, as the higher the price, the lower the level of demand (using that economists' term 'other things being equal'). In the case of demand, building product and service examples include acquiring land, recruiting building workers and using capital goods such as machinery on site. When economists speak of demand they mean something called 'effective demand'. Effective demand is money-backed desire and is distinct from need; therefore it is a demand that can happen because the 'demander' has the resources to pay for it. The determination of demand for goods and services produced by the building industry is a very complicated process, thanks partly to the size, cost, longevity and investment nature of the products and partly to the broad range of what constitutes building activity.

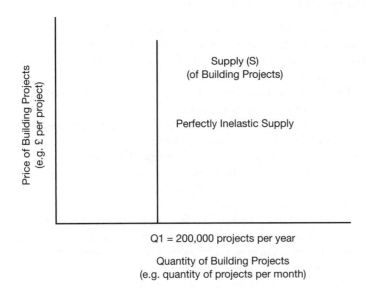

1.4 Perfectly inelastic supply of building projects

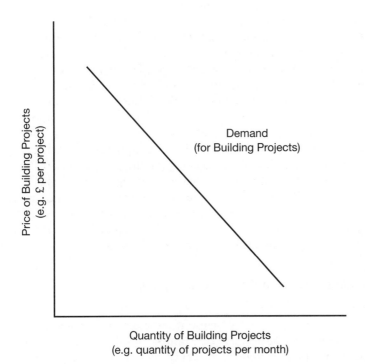

1.5 Demand for building projects

1.5.5 Non-price determinants shifting the demand curve

Changes in non-price determinants will cause this demand curve to shift to the right or to the left, and demonstrate the fact that more or less is being demanded at each and every price. These changes are often referred to as *increases* or *decreases* in demand. It is particularly important to remember the distinction between a movement along and a shift in a demand curve. These rules will not only help us to understand the graphical analysis, but they will also enable us to acknowledge the numerous factors that come into play when interpreting demand in the building industry. There are many non-price determinants of demand, such as the cost of financing (interest rates), technological developments, demographic makeup, the season of the year and fashion. Four major non-price determinants of demand will be briefly focused on here to demonstrate how demand is played out in building projects. These determinants are: income; price of other goods; expectations; and government policy.

First, an increase in income will lead to a rightward shift in the position of the demand curve. Goods where the demand increases when income rises are called normal goods. Most goods are 'normal' in this sense. There are a small number of goods for which demand decreases as incomes increase: these are called inferior goods. For example, the demand for private rented accommodation falls as more people become able to buy their own homes. Second, the prices of other goods act as a non-price determinant of demand. Demand curves are always plotted on the assumption that the prices of all other commodities are held constant. It may be the case that the other good is a substitute good or a complementary good. In considering these substitutes a change in the price of an interdependent good may affect the demand for a related commodity.

Expectations are a third key non-price determinant of demand. For instance consumers' views on the future trends of incomes, interest rates and product availability may affect demand. The potential house purchasers who believe that mortgage rates are likely to rise may buy less property at current prices. The demand curve for houses will shift to the left, reflecting the fact that the quantity of properties demanded for purchase at each and every price has reduced as a result of consumer expectations that mortgage rates will rise. As was the case in non-price determinants of supply, government decisions and current policy affect the demand for a commodity. For example, changes in building regulations may increase the demand for energy-efficient boilers, regardless of their present price. The demand curve for energy-efficient boilers will shift to the right, reflecting the fact that greater quantities of these units are being demanded at each and every price. If a non-price determinant of demand changes, we can show its effect by moving the entire curve from D to D_1, as shown in Figure 1.6. Therefore, at each and every price, a larger quantity would be demanded than before. For example, at price P the quantity demanded increases from Q to Q_1.

1.5.6 Equilibrium in building projects to determine the level of procurement

In considering demand and supply, we have confined our discussion to isolated parts of the market relating to the consumer or producer. Obviously, this separation is theoretical and only useful for educational purposes. In reality, there is a very close relationship between the forces of demand and supply. The interaction of supply and demand that determines prices is seen to gravitate to a point at equilibrium (or market

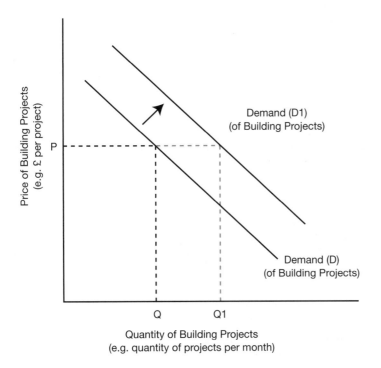

1.6 A shift in the demand for building projects

price) at which both consumers' and producers' wishes are met. The intersection of supply and demand at equilibrium, along with the corresponding market price and quantity, is demonstrated in Figure 1.7. Knowing and understanding how supply and demand interact is an essential prerequisite for interpreting many markets, including building- and construction-related markets.

1.6 Summary and tutorial questions

This chapter has enabled you to understand and apply introductory knowledge of building procurement. Furthermore, the chapter has allowed you to set procurement in the wider context and structure of the building industry. It has also provided detailed stages in the building-procurement cycle, whilst introducing concepts of value for money within a market system. This deeper understanding of markets via supply and demand has demonstrated how changes in prices and quantity can be influenced via differing (and often opposing) forces that tend towards equilibrium.. Market change from non-price determinants has also been demonstrated, showing how, for example, government can influence both supply and demand. Distortions in the building market have been shown that result from the peculiarities of the building industry, especially in relation to building projects often having a more inelastic supply. Other forces are discussed in the chapters that follow, which provide a solid grounding in what is critical to building procurement. This chapter concludes with listed summary points

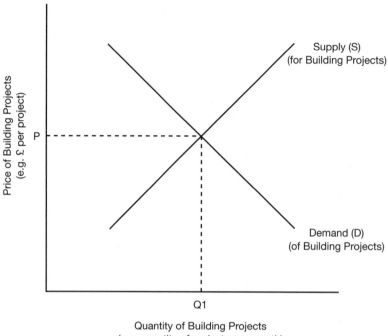

1.7 Market equilibrium in the supply and demand for building projects

and a set of tutorial questions that will enable you to think more deeply about the
building-procurement issues introduced.

1.6.1 Summary

Introduction and nature of the building industry

- The building sector is investment-led, especially if procurement of a building buys
  an 'asset' which has the ability to generate funds in the future.
- Cycles in building activity show far greater amplitude than equivalent cycles in
  general business activity.
- The building industry covers a wide range of business activities that are brought
  together by a common interest in the development of land and real estate.
- Building procurement is frequently complicated because of the unknown variables
  at the start of the project such as the quality of the job, the time it will take and
  the cost involved.

The structure of the building industry

- Since the mid-1990s in the UK, Greater London, the south-east of England and
  the east of England have had growth in output.

- The UK building industry, as it is geographically removed from mainland Europe, does not experience the full influence of cross-border competition on its domestic market.
- Property assets held by BPF members are currently worth over £70 billion and BPF members liaise closely with government via the federation's building committee.

Stages in a building project

- Key building project stages can include pre-design, design, bidding/negotiating, building, operation, delay and abandonment.
- Stakeholders operating within these stages include financial organisations, government agencies, engineers, architects, insurance companies, contractors, material manufacturers and suppliers and building tradesmen.

Value for money in competitive markets

- The level of trade-off between time, cost and quality will determine a project's value for money in a competitive market.
- Consideration of the interrelation of sub-elements of time, cost and quality are critical to ensuring value for money is gained as part of the procurement process.

Supply side and demand side in building procurement

- The forces of supply and demand bring together producers and consumers in a market that operates in such a way that appropriate goods and services are produced, and appropriate incomes are awarded.
- Building supply includes land as a site of building, labour in the building sector such as tradesmen and capital available in the form of machinery and finance.
- The number of potential contractors interested in bidding to supply a project will increase as the profit margin the client is prepared to meet rises.
- Almost half of all building activity is by small general builders dealing with repair and maintenance contracts.
- Non-price determinants of supply increase both price and quantity (demonstrated as a shift in the supply curve). Examples could include an increase in the costs of production, technology, government policy, expectations and supply-chain management.
- In the short run supply of property is inelastic, meaning that adjustment in increasing the supply of property is slow and therefore rental values and house prices are demand-determined.
- The demand for goods and services produced by the building industry is complicated partly because of the size, cost, longevity and investment nature of the products.
- Four major non-price determinants of demand (demonstrated by a shift in the demand curve) are income, price of other goods, expectations and government policy. Other non-price determinants of demand include the cost of financing (interest rates), technological developments, demographic makeup, the season of the year and fashion.

1.6.2 Tutorial questions

1   Essay question: What is the nature and structure of the building and construction industry in the UK?
2   Using the example of a mixed-use property development project, outline all of the time, cost and quality elements. Form a detailed conclusion using percentages as to what element(s) of time, cost and quality are most significant in this development.
3   Give examples in the building sector of non-price determinants in both supply and demand. Using demand and supply analysis diagrammatically show what would happen to a housing market if the majority of stock was demolished to be replaced with fewer high-quality houses that were in high demand.

2 Clients of the construction industry

2.1 Types of clients

All construction projects must begin with the client. This is the party who has instigated the project, will have thought about why the facility should be built, will have organised the funding and be convinced that it is a worthwhile investment. They are therefore the sponsor of the whole construction process, who provides the most important perspective on project performance and whose needs must be met by the project team. However, the term 'client' implies that it is one person or one organisation to whom all other parties can refer, which is not necessarily the case, especially with large, complex organisations where users, decision-makers, financers may all work in quite separate departments, each with its own procedures, priorities and attitudes. Additionally, clients (or customers/consumers) of any industry are not a homogeneous group, and it follows that different clients, or categories of clients, will require different solutions for their requirements and will present different opportunities and challenges for the suppliers of the services.

Therefore we must ensure that, before addressing the technical, managerial and design aspects of the project, the identity, nature and characteristics of the client are comprehensively and accurately identified and that the project team is fully aware of, and understands, the client's particular needs. This is carried out in the briefing stage (see RIBA *Plan of Work* in Chapter 6) and is essential in order to avoid misunderstandings later in the project which may result in lengthy and costly disputes between the client and the design and/or construction team.

A reasonable simple definition of the client is therefore:

The organisation, or individual, who commissions the activities necessary to implement and complete a project in order to satisfy their needs and then enters into contracts with the commissioned parties.

We must remember that the client commissions the building or facility, but is generally not familiar with the workings of the construction industry. This is not unusual, since we all buy cars or other consumer goods without knowing how the industry that produces them is organised. However, as mentioned in Chapter 1, the construction industry is different in that there has traditionally been a separation between the design of the product and its manufacture, which is not the case in the motor car industry; for example, the Ford Motor Company both designs and produces its cars.

Nor is it the case in many other industries. Additionally, if you wished to purchase a Ford motor car, it is because you have seen one on the road, or even had a test drive yourself. Clearly, this is not possible for construction projects as each project is designed and produced separately, on different sites, at different times and to different client requirements. They are all one-off products designed to satisfy the bespoke needs of the particular client at a particular time

Because clients may not be familiar with the workings of the industry, they must rely on their designers to produce an acceptable design and on the builders to build it so that it satisfies their requirements when they have moved in. For this reason, all parties have a formal contract with the client, in order to be clear about their responsibilities to the client as well as to each other. The various contracts will cover the design (drawings, specifications), cost (bills of quantities or activity schedules) and time (date for commencement and date for completion). This is the beginning of the time–cost–quality triangle as discussed in Chapter 1.

The importance of identifying the real client has been amply demonstrated in both theory and practice. Strange as it may seem, problems often occur because the 'real' client may not be the person/company who initiates the project, as we shall see.

As well as individual client organisations being complex and disparate, modern construction projects increasingly use consortia and partnerships of developers which are brought together for the funding, construction and operation of facilities. Clearly, this can also exacerbate the difficulties of communication.

Problems of this sort are most likely to be experienced when dealing with many-faceted organisations in which large numbers of departments and/or individual managers all have their own priorities. Many developers are not the end-users who will eventually occupy the facility. In this situation, difficulties will stem from conflicting requirements and expectations, and it is therefore essential that these are resolved during the early stages of the design and certainly before appointing contractors. The appointment of a single representative to co-ordinate clients' varying requirements will at least reduce the negative consequences of conflicting needs. These single representatives are called the employer's agent or client's representative depending on the procurement process which is adopted.

In the same way, a situation in which the future occupier of a facility is obtaining funding from an external source, or is a tenant of the client, raises the likelihood of similar conflicts arising.

Establishing the identity of the real client is not always easy, and reconciling the demands of different end-users, both within and outside the client's organisation, always requires considerable expertise, ingenuity, tact and diplomacy. However, the increasing complexity of modern client organisations, and the complicated structures of most consortia, especially those set up for PFI and PPP projects (see Chapter 11), means that the industry must accept the need to deal with these issues as a matter of course.

2.1.1 *The categorisation of clients*

Construction clients can be categorised as follows:

1   Whether the organisation is publicly or privately owned or funded (i.e. is from the public sector such as central or local government or from the private sector such as a plc or limited company).

2   The level of knowledge and experience within the client organisation in dealing with the construction industry and implementing building projects (generally referred to as 'sophisticated' clients or 'non-sophisticated' clients). The term 'unsophisticated' would not be considered a complimentary expression by those who are paying the bills!
3   Whether the project is needed by the client to accommodate their own industrial or commercial activities or whether they are acting as a developer to pass on to others, by either leasing or selling outright.
4   The degree of contribution that the client is making to the actual project management of the facility.

A detailed examination of these four characteristics reveals a number of key issues.

Ownership structure

Clients of the construction industry have traditionally been categorised as 'public sector' or 'private sector' as a reflection of the ownership or source of funding of the organisation. However, this is not sufficient in the modern world of small government, quangos and the private finance initiative (PFI). A further third category can be made, that of 'not-for-profit' organisations, e.g. housing associations, regional development organisations, etc. The developments that these clients procure were traditionally seen as public facilities (social housing, local infrastructure and so forth) but have now been handed over by the government to commercially run organisations who are legally bound only to financially break even, i.e. not to make a profit, while still encompassing the good business practices of the private sector.

The characteristics of these categories of client differ mainly as a result of the source of their income, with publicly financed bodies having a requirement to account for taxpayers' money. Internal regulations, standing orders and continuous control and auditing of expenditure are also used to ensure public accountability, and each of these safeguards can have the effect of limiting the choice of procurement procedure. The accounting structure of these organisations therefore often concentrates solely on expenditure control as income is generally determined by others.

Not-for-profit organisations have a much greater degree of flexibility in their management and investment policies, while still being accountable to government for their overall performance.

Privately owned organisations are concerned with maximising profits and shareholder value and are therefore prepared to adopt a more risk-taking approach to procurement. After all, basic economic theory suggests that profit is the reward for taking risks. Their accounting structures therefore concentrate on both income maximisation and expenditure control.

Level of experience

The level of the client's experience of the construction industry and project implementation is also a critical characteristic in terms of the relationship between the client and the supply-side partners in the construction industry. The client's attitude to all aspects of construction activity is determined by whether they are 'experienced' or 'inexperienced'.

Experienced clients, such as government departments, large private companies, supermarket chains, etc. normally have a department which deals with their construction and property portfolio and will therefore generally have:

- a detailed knowledge and understanding of the construction industry and its procedures;
- a continuing, or regular, involvement with the construction industry, including at times the implementation of large high-value and complex projects;
- the ability to produce a comprehensive initial brief incorporating prioritised objectives for the cost, timing, quality and functionality of the project;
- expertise in the overall management and control of construction projects and of construction consultants and contractors;
- the employment of in-house construction managers and sometimes designers of various construction disciplines;
- a desire to be constructively, consistently and continuously involved during the life of the project without detriment to the powers, responsibilities and status of the appointed consultants and/or contractors.

Inexperienced clients, on the other hand, are irregular or one-off clients and therefore have:

- a lack of continuing or regular involvement with, or knowledge of, the workings of the construction industry or the implementation of construction projects other than in a minor capacity;
- a lack of expertise in the overall management of construction projects and/or construction consultants or contractors;
- a lack of understanding of the importance of the early production of an initial brief and a set of prioritised objectives, without substantial assistance from an external construction consultant;
- a limited understanding that client-imposed changes after the commitment to construct stage has been reached will have consequences for both the cost of the project and possibly its time duration;
- the potential to be influenced on construction matters by external parties.

Primary or secondary business activity

The third characteristic relates to the reason for the client's need to build, i.e. whether the facility is required for their own separate business activities or whether building is their primary business activity which is then leased or sold on to others:

- primary clients (developers) whose main activity and primary source of income derives from constructing buildings for sale, lease, investment, etc.;
- secondary clients who require buildings to enable them to house and undertake their own main business activities and whose expenditure on construction represents a small proportion of their annual turnover.

The vast majority of primary clients will, by definition, be experienced and largely involved in public or private property development. Secondary clients will consist

of a wide range of large, medium and small manufacturing and service organisations within the industrial and commercial sectors and will exhibit various levels of
experience.

This typology is the main way of obtaining information from clients of the construction industry to determine changes in output and procurement strategies.

Figure 2.1 offers a graphical representation of these categories. In terms of the
implementation of construction projects, and particularly the selection of procurement methods, it needs to be borne in mind that subcategories of all the listed categories probably exist, but for our purposes this representation is sufficient.

Statistics relating to construction output, prepared by both government departments
and the industry itself, categorise the output in terms of newbuild or refurbishment
and whether the client is a public or a private client, etc. (see Figure 2.1).

As Table 2.1 overleaf shows, the total value of the UK construction industry by
the end of 2008 (at 2005 prices) was £109.705 billion and as the total GDP of the
UK in 2005 was approximately £1,350 billion, the construction industry therefore
represents just over 8 per cent of activity in the UK economy, as mentioned at the
beginning of Chapter 1. This total figure is made up of £47.224 billion (43 per cent)
for repair and maintenance of existing buildings and £62.481 billion (57 per cent) for
new construction works including both building and civil engineering infrastructure.

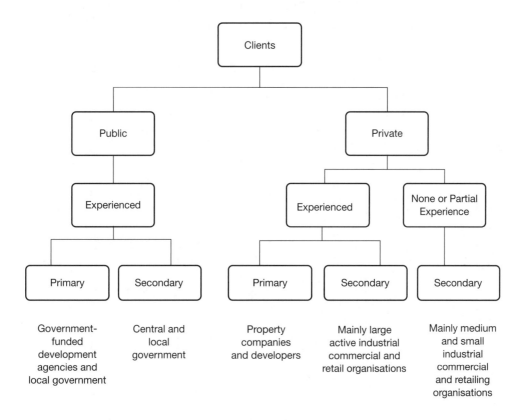

2.1 Categories of clients in the construction industry

Table 2.1 Example of construction output data

Table 3 Volume of construction output by all agencies (1) by type of work (at Constant (2005) Prices)

Great Britain

£ Million

| | New housing | | Other new work | | | | All new work | Repair and maintenance | | | | All repair and maintenance | All work |
| | Public | Private | Infrastructure | Excluding infrastructure | | | | Housing | | Other work | | | |
				Public	Private Industrial	Private Commercial		Public	Private	Public	Private		
2002	2,163	13,760	9,053	8,355	3,429	17,103	53,861	7,561	14,850	7,859	14,389	44,658	98,520
2003	2,384	15,652	8,217	10,120	3,669	16,297	56,340	8,112	15,563	8,818	15,181	47,674	104,013
2004	2,884	17,723	7,133	11,197	4,036	17,191	60,163	8,766	15,787	8,473	14,663	47,689	107,852
2005	2,680	18,383	6,499	10,191	4,291	17,369	59,412	8,598	15,339	8,939	14,718	47,594	107,006
2006	3,281	18,714	6,008	9,679	4,767	19,695	62,145	8,352	14,858	8,276	14,733	46,219	108,364
2007	3,778	18,410	6,189	9,205	4,785	22,178	64,544	7,987	15,054	7,398	15,968	46,407	110,952
2008 (R)	3,480	14,845	7,120	10,646	3,861	22,529	62,481	8,170	15,507	8,106	15,440	47,224	109,705
2003 1	575	3,711	2,072	2,216	826	4,029	13,428	1,786	3,687	2,267	3,591	11,332	24,760
2	607	3,809	2,083	2,405	856	4,021	13,782	2,063	3,899	2,136	3,710	11,808	25,590
3	599	3,941	2,082	2,627	947	4,043	14,239	2,188	3,906	2,199	4,054	12,346	26,585
4	602	4,191	1,980	2,873	1,040	4,205	14,890	2,074	4,071	2,217	3,826	12,188	27,078
2004 1	731	4,341	1,847	2,858	1,010	4,230	15,016	2,380	4,031	2,242	3,839	12,492	27,508
2	767	4,415	1,867	2,851	995	4,377	15,272	2,105	3,913	2,030	3,589	11,636	26,908
3	724	4,517	1,821	2,775	999	4,319	15,155	2,096	3,941	2,078	3,598	11,713	26,868
4	663	4,450	1,598	2,713	1,032	4,264	14,720	2,185	3,902	2,124	3,637	11,847	26,568

Year	Quarter													
2005	1	691	4,431	1,618	2,625	1,007	4,193	14,564	2,409	3,800	2,313	3,694	12,216	26,780
	2	676	4,676	1,603	2,561	1,059	4,331	14,907	2,276	3,867	2,243	3,689	12,075	26,981
	3	638	4,681	1,653	2,486	1,087	4,367	14,913	1,990	3,796	2,210	3,726	11,721	26,635
	4	674	4,595	1,625	2,519	1,138	4,477	15,028	1,923	3,877	2,173	3,609	11,582	26,610
2006	1	798	4,569	1,599	2,525	1,181	4,642	15,313	2,145	3,744	2,126	3,554	11,569	26,882
	2	844	4,652	1,495	2,422	1,162	4,771	15,344	2,036	3,808	2,121	3,646	11,611	26,955
	3	839	4,757	1,476	2,378	1,179	5,067	15,695	2,148	3,603	2,099	3,619	11,470	27,165
	4	801	4,736	1,439	2,355	1,246	5,216	15,793	2,022	3,703	1,930	3,913	11,568	27,361
2007	1	960	4,672	1,400	2,301	1,251	5,280	15,864	2,149	3,642	1,840	3,952	11,583	27,447
	2	980	4,677	1,538	2,261	1,237	5,452	16,145	1,968	3,814	1,836	3,908	11,526	27,671
	3	948	4,642	1,624	2,316	1,170	5,680	16,380	1,922	3,616	1,853	4,035	11,426	27,806
	4	890	4,419	1,626	2,329	1,127	5,765	16,156	1,948	3,983	1,869	4,073	11,873	28,029
2008	1	915	4,226	1,749	2,508	1,132	5,964	16,494	2,031	3,749	2,013	4,049	11,841	28,335
	2	894	3,870	1,859	2,586	961	5,637	15,807	2,117	3,981	2,031	4,077	12,206	28,013
	3	883	3,593	1,876	2,756	927	5,798	15,833	2,083	3,723	2,147	3,807	11,760	27,594
	4	788	3,156	1,635	2,796	841	5,130	14,346	1,939	4,055	1,915	3,508	11,416	25,763
2009	1 (P)	790	2,845	1,774	2,951	683	4,443	13,486	1,933	3,389	1,879	3,305	10,506	23,992
	2 (P)	823	2,823	1,995	3,270	590	4,060	13,561	1,927	3,388	1,779	3,220	10,315	23,876

Source: UK Government Office for National Statistics 2010.

Notes

1 Output by contractors (including estimates of unrecorded output by small firms and self-employed workers) and output by public sector direct labour departments classified to construction in the Revised 2003 Standard Industrial Classification.

P = provisional; R = revised

The 2009 figures are given as provisional, and it will be interesting to see how far these are affected by the world economic downturn. Private commercial buildings (shops, offices, etc.) represent the largest sector, followed by private residential construction. Public building work (schools, hospitals, etc.) is at about half the level of private commercial building.

Other methods of classifying clients do also exist, but they tend to be a development of the above rather than a completely new form of classification. For our purposes in this introductory book, the above description is considered sufficient.

No categorisation system will ever succeed in totally capturing the many variants and subspecies of its subject matter. Clients of the construction industry are no different in this respect from any other group of individuals or organisations; some will always exhibit non-stereotypical characteristics and remain unclassified or even unclassifiable. In the main, however, it is suggested that the majority will remain remarkably faithful to the attributes that have been identified, although care must always be taken to ensure that the specific characteristics of each organisation or individual are established as a first priority.

2.2 Clients' objectives and needs

The client's objectives and needs clearly have to be central to any procurement decisions, in the same way that as a customer we all go to the shops to buy, say, a pair of shoes. Why do we need the shoes? For work? For leisure? As a fashion statement? Do the shoes need to be hard-wearing (for hill walking) or light and fashionable (for socialising)? We all make these procurement decisions all of the time, and the answers we come up with are derived from our objectives and needs at the time.

The Latham Report (HMSO 1994a) was produced in 1994 on behalf of the UK government, which is a major construction client, and the report suggests that a client's project needs are:

- obtaining value for money;
- ensuring the project is delivered on time;
- having satisfactory durability;
- incurring reasonable running costs;
- being fit for its purpose;
- being free from defects on completion;
- having an aesthetically pleasing appearance;
- being supported by meaningful guarantees.

The Egan Report in 1998 (Department of Trade and Industry 1998) went even further as it was largely written by major clients and focused on how the construction industry should change to meet the developing requirements of these clients. Both of these reports are seminal works of the modern construction industry and will be referred to constantly throughout the remaining chapters of this book.

There is little doubt that the requirements mentioned above reflect the needs of most clients, either in whole or part, but let us now look at the client's needs and objectives that are particularly related to the project procurement process to see if they can be prioritised for different categories of client.

Whatever the individual client's needs and objectives, it is essential that both the client's characteristics and their objectives are identified and understood by the project team as quickly as possible. As mentioned previously, this should be carried out during the preliminary briefing stage of the project and, as time spent in preparation is never wasted, this is arguably the most important stage of the entire project as one which sets the foundation for any ambiguities or disputes later on. To take away ambiguities at the beginning of a project has the very positive effect of reducing the possibility of disputes later on.

Prior to the publication of the Latham Report, many detailed studies of the needs and objectives of clients had been carried out. For example, in 1975, the Wood Report (National Economic Development Office 1975) revealed the results of case studies of fifty building and civil engineering projects which showed that clients consistently mentioned the need to meet the criteria of cost, low maintenance charges, time, quality, functionality and aesthetics as being necessary for a project to be considered successful. Bennett and Flanagan's (1983) Reading University report 'For the Good of the Client' prioritised the client's requirements as:

- a functional building, at the right price;
- quality, at the right price;
- speedy construction;
- a balance between capital expenditure and long-term ownership costs;
- identification of risks and uncertainty associated with the project;
- accountability, particularly in the public sector;
- innovative design/high-technology buildings;
- maximisation of taxation benefits;
- flexibility to enable design to be changed;
- building which reflects the client's activities and image;
- minimisation of future maintenance;
- the ability to keep any existing buildings operational during the construction period;
- an involvement in, and a need to be kept informed about, the project throughout its life.

This list was never intended to be exhaustive or to suggest that all clients would require all the points in every project. Rather, it is suggested that the list reflects most of the elements of most clients.

In the same year that this list was produced, the landmark report *Faster Building for Industry* was published by the UK government's National Economic Development Office (NEDO) (National Economic Development Office 1983). The results were based upon research to establish clients' attitudes to timing and speed in their projects, and concluded that few clients were interested in speed for its own sake. The report also found that very little consideration had been given to the influence of time on the financial aspects of projects, with clients, in general, being prepared to incur additional costs in order to achieve faster construction times and being uncertain about the role of time in the construction process.

All of these reports, together with several subsequent studies, have confirmed that clients wanted certainty of performance in all three of the major criteria of time, cost and quality. They did not want any surprises during the implementation of their projects and specifically required:

- value for money;
- a durable and easily maintained building with affordable running costs, no latent defects and rapid rectification of any minor problems that may occur;
- clear allocation of responsibilities among the members of the project team;
- a minimal exposure to risk;
- an early indication of a firm price for the project and comprehensive information on any future contractual claims;
- an early start on construction work with minimal interference from external sources such as planning or building control;
- a non-confrontational business relationship with the contractor, who should give guarantees and good 'after-sales' service.

Each client will obviously have different specific objectives for each of their projects, but the lists above should act as a useful classification of what the client wants. If these are considered carefully at the beginning of the project and kept in mind throughout the various stages of the project, the client will be more likely to be satisfied with the end product. A satisfied client is much more likely to return to your organisation with their next project.

2.3 Project constraints

It is all very well construction clients having these various objectives, needs, requirements and so on, but the supply side of the industry (i.e. the contractors, subcontractors and suppliers) does not work in a totally controllable environment. The constraints on the supply side must therefore be taken into account when considering how to satisfy the client's objectives most optimally (i.e. the procurement strategy).

The effect of these factors will be to influence or restrict the achievement of the client's objectives. The constraints may be classified as:

- *physical*, for example location of the site, difficult ground conditions, high water tables, etc.;
- *organisational*, for example constraints due to the decisions on method statement, programme, etc.;
- *legal*, for example health and safety considerations;
- *economic*, for example a fiercely competitive tender price or guaranteed maximum price for a design build project;
- *political*, e.g. planning authority requirements or particularly sensitive local neighbours;
- consequence of the client organisation's policies, culture or internal regulations.

The constraints stemming from internal rules would also restrict the actions of construction managers in order to safeguard the interests of shareholders or taxpayers. For example, the public client may only be able to accept tenders which enable them to enter into a type of contract which ensures that a lump-sum price is agreed with the most acceptable bidder and that it is fixed for the duration of the project. Such an approach would eliminate other possible procurement methods.

When initially examining a potential project, all of its constraining factors need to be identified. At the same time, all of those parties within the client's organisation having an interest in the development, together with any external forces which may be able to exert any restraining influence, will also need to be considered and factored into the decision process.

As mentioned, these factors will be assessed at the preliminary briefing stage, if necessary using one of the number of techniques now available for this purpose, in order to quantify their actual constraining effect on the project objectives and the procurement process.

2.4 Risk and value

The management of risk is a wide-ranging and mature subject which is of vital importance in ensuring the successful implementation of the project and particularly the selection of the most appropriate procurement method, each of which carries a different level of risk for the main participants. One of the primary purposes of procurement route selection is to allocate the risks between the parties to a project. At one end of the scale, the client can accept all the risks if they carry out the project using their own resources, while at the other end, the risks are transferred to the contractor under package deal or turnkey arrangements. By transferring risk, the client also loses control, so these decisions have to be made carefully.

Risk analysis and management is a very complex issue, and this section can provide only a general introduction to the principles in order to clarify its relation to the procurement process. Risk is effectively about uncertainty; by placing a bet on a horse, you are uncertain whether it is going to win, therefore you are risking your bet. However, if it does win, you will receive more money back, depending on the chances (odds) of it winning. Clearly, overall not everyone can be a winner, except of course the bookmaker.

Risks in construction projects can be classified according to the following general groups:
Risks associated with economic activity (both macro and micro):

- project funding, interest rates and currency fluctuations
- market conditions at the time

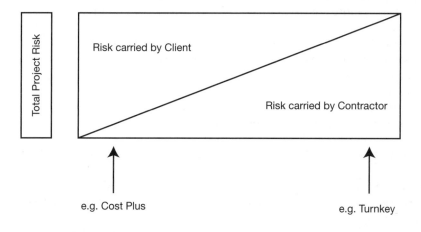

2.2 Allocation of risk between client and contractor

- quality of estimating and tendering data
- unrealistic client business plan for the project
- financial stability of the contractor and other members of the supply chain.

Risks associated with political situations:

- change of central government with possible changes in policy direction
- change of local government with possible change in policy direction (e.g. planning).

Risks associated with physical conditions of the site and location:

- nature of the site (e.g. topography, ease of access, restrictions of space, security, etc.)
- location of the site (e.g. distance from road network and other services, rural versus urban)
- weather conditions (including time of year for proposed project)
- *force majeure* (these are defined as situations which neither party could have foreseen).

Risks associated with legal and contractual decisions:

- contractual allocation of risk between client and contractor
- business failure of project participant (e.g. if the main contractor becomes insolvent, what happens to the subcontractors?).

Risks associated with the organisation of design and construction:

- client organisation – which department of the client is responsible for project liaison?
- consultants – previous experience and competence
- nature of design solution – does it work?
- complexity of design – is the building over-designed?
- design completion before construction – very rarely achieved and therefore a risk for the contractor
- methods of construction – tried and tested construction methods?
- health and safety implications – all risk assessments carried out on procedures?
- programme and duration – are they achievable?
- quality of resources – sufficient for the project?
- unforeseen ground conditions – is there a recent ground-condition survey?
- security – potential for loss owing to theft or damage.

In order to manage these risks effectively, a four-stage process should be undertaken:

1 identification of the sources of potential risk using the lists above (risk identification)
2 analysing the potential effects of each risk item if it occurred (risk analysis/ evaluation)
3 development of a form of response to mitigate the identified risks (risk response)
4 providing for the risks that are left over (residual risk).

2.4.1 Risk identification

The different elements of risk are generally identified in a risk log or risk register, which should identify the sources of risk, the discrete characteristics of the project and which elements of risk are likely to endanger a successful outcome to the project.

The Office of Government Commerce in the UK (www.ogc.gov.uk) states that the purpose of a risk log is to 'identify the risks and the results of their evaluation. ... These details can then be used to track and monitor their successful management as part of their activity to deliver the required, anticipated benefits.'

The OGC suggests the following contents for a risk log/register:

- risk identification number
- risk type (where indication helps in planning response)
- risk owner raised by (person)
- date identified
- date last updated
- description
- cost if it materialises
- probability
- impact
- proximity
- possible response actions
- chosen action
- target date
- action owner/custodian (if different from risk owner)
- closure date
- cross-reference to plans and associated risks, which may also include
- risk status and risk-action status.

Issues and risks can be raised by anyone involved in the project or its stakeholders throughout its life cycle.

Where subcontractors, suppliers or other project partners are involved, it is essential to have a shared understanding of project risks including any agreed plans to manage them. The risk log/register is created during the start-up of the project to be ready to record project risks, including any noted in the project briefing from the client.

The risk log/register is a management tool to be used for a process of identifying, assessing and managing risk to acceptable levels. It is intended to provide a framework in which problems and hazards which may adversely affect the delivery of the project are captured and actions instigated to reduce the probability and impact of the particular risk item.

2.4.2 Risk analysis/evaluation

The effects of the risks, once they have been identified, need to be quantified by means of risk analysis, with the choice of the analysis technique being determined by the potential effect of the risk factor if it occurred, the level of expertise and experience of the project team and the size and complexity of the project.

The most basic form of risk analysis will assess the financial or time effect of the risk (or hazard) as major, intermediate (i.e. some effect on cost and time) or insignificant. Additionally, the probability of occurrence will also be assessed, which will give some indication of the actual financial effect of the risk item. Table 2.2 illustrates this point.

The table uses what is known as a 'stochastic' distribution of probabilities, in that there are only three possibilities – a major effect, some effect or a minor effect on the project. More complex risk-management systems would use a 'continuous' distribution (i.e. the standard 'bell' diagram of a normal distribution curve), which gives a more accurate distribution of probabilities and interdependences. The analysis, which is well outside the scope of this book, would have literally millions of permutations and calculations, so would need to be carried out by computer program to simulate the effect on the project of the different probabilities of occurrence. One of the computerised methods is called the 'Monte Carlo simulation' and implies yet another link between risk management and gambling. But then some may argue that effective gambling is merely an applied form of probability analysis – i.e. risk analysis.

2.4.3 Risk response and mitigation

The risks that must be dealt with are those where the effect multiplied by the likelihood of occurrence create a potential loss which the client is unwilling to take a chance on. The ways of dealing with these risks are:

a   Avoid the risk – by management action such as changing the method of work.
b   Reduce the risk – for example by obtaining more information so that the level of uncertainty is reduced or by taking preventative or protective action. The use of PPE (personal protective equipment) by workers on site is an example of this.
c   Transfer the risk – by taking out insurance or contractually transferring it to others.
d   Retain the risk – take the chance yourself. If the hazard occurs – you pay.

Clearly, there will be a cost in some of these response strategies, and it will be up to the risk manager to decide if the cost of the response is appropriate to the likelihood of the risk.

Table 2.2 Example of a simple risk register

Risk item	Effect if occurred	Possibility of occurrence			Actual effect
	Major/Some/Minor	Likely	Maybe	Unlikely	
1	Major	Yes			Major effect on the project – needs to be dealt with.
2	Some			Yes	Unlikely to have any effect – can be ignored.
3	Minor		Yes		Unlikely to have any effect – can be ignored.

Retaining the risk is, in most cases, likely to be the last resort, as many clients are risk-averse and keen to allocate any potential problems to others. Where responsibility is retained, an allowance is usually included within the project budget as a contingency fund. Insurance cover is often legally required for third-party injuries and can be obtained for some well-defined specific risks, but may not be cost-effective. Whichever of these four routes is taken, the principle of allocating the risk to the party most capable of dealing with it at the minimum cost to the project should always be followed.

If it is decided to retain the risk within the company, there are some simple rules to make sure that the risk element is not forgotten.

- Put it on the agenda for regular meetings so that it does not get forgotten about.
- Incentivise the team to regularly address the risks.
- Make the risks pertinent to the project and its current stage.
- Make sure that progress in mitigating risks is highlighted so that the team feel that they are getting somewhere.
- Make sure that there are some tangible outputs from the mitigation strategies selected.

That now completes our discussion of the construction client, i.e. the demand side of the industry, following the discussion in Chapter 1, which formed an introduction to the supply side of the industry.

The remaining chapters are devoted to examining the methods of procuring the project and the advantages and disadvantages of each.

2.5 Summary and tutorial questions

2.5.1 Summary

Clients are clearly central to the whole process. They are the purchasers, the customers, the consumers and represent the 'demand side' of the industry, whereas the contractors and consultants represent the 'supply side'.

Clients can be categorised in terms of whether they are publicly funded (e.g. government departments) or privately funded as independent commercial organisations. They can also be categorised in terms of their level of experience as clients of construction work. First, commercial property developers have a primary business focus in the construction and property industries and will either sell on or lease out the building after construction. Second, organisations such as supermarkets and retail chains clearly have a different primary business focus, but need to engage with the construction industry on a regular basis, and the value on the balance sheet of their property portfolio means that they need to employ specific professionals to manage the assets. Third, there are the one-off or occasional clients, whose business may well be in the manufacturing or service industry but will need the construction industry for that extension to the factory, new office, etc. Each of these different types of clients requires a different approach from the supply side.

As with any purchase decision by a customer, the client will have objectives: what do they want out of the process; what do they think they are buying? It is absolutely essential for the supply side to understand what the client wants and to deliver that product as closely as possible. This is actually basic business marketing, and as we all know, if you are happy with a purchase, you will tend to return to that provider again and also tell your friends.

Unfortunately, the purchasing decisions in construction are not that straightforward. As we shall see, all construction projects are unique, and successful procurement relies on an effective allocation of risk between the parties and for the client to feel they have obtained good value from the transaction. Therefore the process of risk analysis and management is an important first step for any client to take at the beginning of a project.

2.5.2 Tutorial questions

1 Why is the client the most important person in the project?
2 What are the five most important objectives of the client in a construction project?
3 List the various categories of construction clients and give examples.
4 How would a construction project be most likely to achieve 'client satisfaction'?
5 Section 2.3 states that the supply side of the industry does not work in a totally controllable environment. What does this mean and how does it affect the contractor's ability to satisfy the client's objectives?

3   Historical development of building procurement systems

Having established the concept of building procurement systems, let us take a look in more detail at the evolution and level of use of procurement methods in the modern era.

The vast majority of the construction projects before the Second World War (i.e. before 1939) were implemented by conventional methods of procurement that had remained virtually unchanged for over 150 years. Since that time, however, the number of different procurement systems available has substantially increased, often imported from other countries and perhaps, more significantly, as a result of the willingness of an increasing number of clients to sponsor and use new methods because of frustration at the construction industry's perceived poor performance.

Four phases in the development of contemporary procurement systems can be broadly identified.

1   A period of sustained economic growth when the use of conventional methods of procurement still prevailed (why change?).
2   A period of recession characterised by a relatively modest increase in the use of non-conventional procurement systems.
3   A time of post-recession recovery during which the most experienced clients introduced new procurement methods, and design and build and management-orientated systems substantially increased their share of the available workload.
4   The period from the early 1990s to the beginning of the 'credit crunch' recession that started in earnest in late 2008. This period contained nearly equal phases of recession and recovery, as well as the advent of partnering, an increase in the use of the various forms of the private finance initiative and efforts by government to improve the industry's performance through the publication of two major reports and changes in the way the industry manages its own projects.

Chronologically, these periods relate to the years 1945–72, 1973–80, 1981–90 and 1991–2008, over half a century during which the changed attitudes and needs of client organisations have done more than any other factor to increase substantially the number and types of available procurement systems.

3.1 Up to 1972

After the end of the Second World War in 1945, the demand placed upon the building industry rapidly increased in terms of both workload and complexity. Despite this, the way in which projects were organised remained largely unaltered, particularly in the public sector, where the greatest part of the rebuilding work was being let on the basis of open competitive tendering, despite the Simon Report of 1944 (Ministry of Works 1944) having strongly recommended the use of selective bidding (tendering).

The Phillips Report published in 1950 reiterated this recommendation and highlighted the need for greater co-operation between all of the parties involved in the construction process (HMSO 1950). By this time, however, some innovative procurement systems, such as negotiated tenders and design and build, had begun to be used on a very limited scale by the private sector and central government.

Criticism of the lack of sound relationships and co-operation between the members of the project team and their mutual clients was contained within the Emmerson Report of 1962, which made the now well-known observation that:

In no other important industry is the responsibility for design so far removed from the responsibility for production. (Emmerson 1962)

Emmerson came to the conclusion that there was still a general failure to adopt enlightened methods of tendering despite the recommendations of earlier reports, but also noted the growth of package deals and 'other forms' of placing and managing contracts. No examples of such forms, other than serial tendering and a passing reference to the CLASP (Consortium of Local Authorities Special Programme) system of industrialised building, were provided.

In 1964, the Banwell Report was published, expressing concern at the failure of the industry and its professions to think and act together or to reform its approach to the organisation of construction projects (Banwell 1964). The report also reiterated the recommendations of the previous Simon Report and other committees and working parties that the use of selective tendering should be more universally applied, together with the use of non-conventional methods of procurement, such as negotiated and serial tendering, where appropriate.

The 1967 review of the Banwell Report, *Action on the Banwell Report*, found that some progress had been made since 1964, specifically on the pre-planning of projects, although the professionals had done little to de-restrict their practices (Economic Development Committee for Building 1967). An increase in the use of selective tendering was noted, and the industry was again urged to increase the use of serial and negotiated tendering. It was also noted that a number of guides published over the decade before 1965, and intended to assist clients in organising and planning their construction projects, had not been sufficiently publicised and circulated, with the result that they had been under-used.

A study in 1965 sponsored by the National Joint Consultative Committee (NJCC) that examined communications in the building industry was probably the first to suggest that overall co-ordination of design and construction should be exercised by a single person or organisation, although it was some years before such a philosophy gained general acceptance (Higgin and Jessop 1965).

The early to mid-1960s saw economic expansion in the UK, with rapidly developing technology, changing social attitudes, the demand for more modern, complex and sophisticated buildings and, not least, the client's increased need for faster completion at minimum cost. All of these factors generated considerable activity within the industry, as a consequence of which the general standard of performance and organisation improved.

The need for cost-effective and faster completion of projects stemmed from the revived activities of private-sector property developers following the rapid increase in post-war urban development and increased activity within the industrial sector. Developers and industrial organisations were not restrained by the standing orders or restrictive procedures necessarily adopted by the public sector and were more open to suggestions for the use of what were then unorthodox arrangements for the provision of their building projects. Negotiated contracts and package deals were frequently entered into by the private sector, and there was much discussion, but much less real progress, about the early involvement of contractors in project planning in order to benefit from their practical expertise in what we now call 'buildability'.

Overall, the period from the end of the war to the early 1970s was that of the 'baby-boomer' generation – a time of sustained and almost uninterrupted economic growth and social progress, during which, in terms of construction procurement, conventional methods prevailed, with only a relatively small number of projects being carried out using non-conventional procedures. This was despite a proliferation of reports recommending their use and advocating the adoption of a more co-operative approach by all members of the project team. This was perhaps due to lack of time to learn new methods and an attitude of 'the current system is working fine, thank you very much'.

3.2 1973–1980

This second phase was a period of recession, if not depression and instability, and commenced as a result of unexpected and large price increases in crude oil imposed by the oil-producing countries coupled with high inflation caused by the previous economic boom.

This stage is seen to last until the end of 1979, although the effects of the oil-price increases in 1973–4 on the economies of Western countries were still being felt during the early 1980s.

During this period, government-sponsored studies of the construction industry tended to be related more specifically to individual sectors of the industry, unlike the more generalised investigations that were carried out in the 1950s and 1960s.

Of the number of studies coming out of the UK's National Economic Development Office (NEDO), the Wood Report of 1975 was the only one specifically to examine purchasing policies and procurement practices, although even then the examination was restricted to the public sector (National Economic Development Office 1975). The report found that the procurement systems used by public authorities were inappropriate to the circumstances of the projects surveyed, although non-conventional methods of procuring construction work were used in 40 per cent of the projects examined.

Various official and unofficial reports were produced during this phase and drew unflattering comparisons between the performances of the UK and a number of overseas construction industries. Two case studies carried out by Slough Estates in 1976 and 1979 were particularly damning.

The 1976 study found that the overall time taken to implement an industrial project in the UK was considerably longer than in other countries, and the eventual cost of developments in the UK was considerably higher than in all but one of the other seven countries surveyed. The reasons for this poor comparative performance were considered to stem from the unnecessarily lengthy and complex design and pricing process (which is undertaken consecutively and before any work is started on site) and the time taken to obtain the necessary statutory permits, which, when related to the high level of interest rates, inflation and prolonged design and construction periods, led to much higher levels of cost for the client.

The 1979 report endorsed the findings of the first report, but conceded that there had been a general improvement in the intervening three years that was attributable to the effects of the economic recession and the resulting low level of building activity, which usually has the effect of improving efficiency as a method of driving down costs in order to create more profit.

The recommendations made by the two studies in the context of the procurement process urged the simplification of design and construction procedures, the improvement of the management of construction, the establishment by clients of their real needs and more effective briefing of the project team at the beginning of the process.

These two reports were probably the first formalised examples of a trend, which emerged during the late 1970s and has continued since, for some major client organisations to forcibly voice their dissatisfaction with the performance of the construction industry.

In 1978, the UK government's Building and Civil Engineering Development Committees combined to produce *Construction for Industrial Recovery*, a study which sought, among other aims, to obtain the views of industrialists on the adequacy of the products and services offered by the construction industry (National Economic Development Office 1978).

The report, based upon a survey of 500 firms and 32 case studies, concluded that industrial firms carried out their construction projects using a variety of procurement methods, with many choosing the traditional route. It was also established that the average industrial user was not aware of the complexities of the construction process, or more particularly the alternative methods of acquiring buildings, and that contract procedures often failed to meet the needs of the manufacturing sector of industry.

The Royal Institution of Chartered Surveyors (RICS) published a report in 1979 which was based on a study by Reading University of the contribution made by different design and contract procedures to a project's cost and time performance, especially in the UK and US construction industries (University of Reading, Department of Construction Management 1979). The document contained a number of conclusions which are relevant to the procurement process:

1   A very wide range of procurement systems is used in the private sector in the USA compared with a very narrow range in the UK, where central and local government is the dominant influence and public accountability, rather than economy, is the overriding criterion.

2   Major time and cost penalties are likely to be incurred if detailed design is separated from construction.

3   The use of construction management in the USA has grown as a result of this system being capable of giving the design team control over the construction time and cost.

4   The range and variety of procurement systems have proliferated on both sides of the Atlantic because of escalating costs, increasingly complex designs, the increased size of projects and the more onerous demands of owners.

5   Clients in both countries were found to be dissatisfied with conventional proce-
    dures. In the USA this dissatisfaction results from the increase in claims, and
    subsequent litigation, as well as a lack of cost control during the design stage.
6   Clients in the UK are more conservative than their US counterparts, who are
    prepared to experiment with the whole range of procurement methods, particu-
    larly – at the time of the survey – construction management.

The theme of most of the reports published in the 1970s reflected this conservatism,
as a diminishing number of clients were prepared to commit themselves to implementing projects in an uncertain economic climate. There was also increasing concern
among those clients who continued to carry out construction works about rapidly
increasing material and labour costs, high inflation and falling demand for their products, all of which were made worse by the delays, overruns and other difficulties associated with the UK construction industry.

Within this second phase, the use of conventional procurement methods still
accounted for much of the construction work, although the use of management
contracting, and to a lesser extent design and build, continued to increase.

3.3 1980–1990

The third of the post-war phases of the UK construction industry started in 1980 and
finished about a decade later. It has been described as a post-recession period of adjustment and recovery, during which major changes took place in the economy as well as
long-term shifts in the structure of the construction industry, such as the emergence of
labour-only subcontracting, and changes in the attitudes of major clients.

This last characteristic is best reflected in the launching, at the end of 1983, of
the British Property Federation's (BPF) *System for Building Design and Construction.*
This body, which represents the majority of UK property development organisations
and a number of large retailers and commercial companies, had concluded that many
existing procurement systems caused delays, were inefficient, increased costs and could
cause and sustain confrontational attitudes between consultants and contractors that
were not in the best interests of the client.

The Federation had set up a working party, assisted by a number of consultants,
to draft 'an improved management system for the building process appropriate
to members of the BPF' and eventually produced its own system of procuring
and implementing building projects. The system reflected US practice and the
experience gained by its members in using all of the various existing procurement methods that had become established over many years. However, although
the new system attracted much comment at the time, this new approach hardly
captured the imagination of many clients or a significant percentage of the available workload.

At the same time, two important government-sponsored reports, *Faster Building for
Industry* and *Faster Building for Commerce*, were published by NEDO in 1983 and
1988. Both dealt with the widely held belief that the current process of procuring new
industrial and commercial buildings was unnecessarily long and difficult. This view
had already been expressed in the reports of the 1970s and also compared the UK
construction industry performance unfavourably with that of a number of overseas
countries.

The 1983 study found, in relation to the procurement methods used on industrial projects, that whereas conventional methods of procurement could give good results the use of non-conventional techniques tended to result in faster progress, although it was also established that the use of approximate bills of quantities and negotiated tendering led to faster project implementation. It was also determined that over half of the projects surveyed were carried out by conventional means, about one-third by the design and build method and the remainder by some form of management or other approach.

The 1988 report, which dealt specifically with commercial projects, identified considerable variability in time performance even on similar projects. Project outcomes had been determined not only by the form of organisation but also by the early development of a comprehensive project strategy and timetable. Two-thirds of the projects surveyed had been carried out using conventional methods, with the remainder being more or less evenly divided between design and build and management methods.

Both studies found that about one-third of all industrial and commercial projects were completed on time, but that the remainder overran their planned times by one month or more, thereby confirming the concern expressed by many clients during this period at the industry's inability to satisfy their basic needs.

The frustration of clients at this inability resulted in the British Property Federation's production of its own system of managing projects. Other major clients, mainly those implementing megaprojects in London and the south-east of England, began in the mid-1980s to individually produce their own forms of procurement in order to meet their own particular objectives and interests more efficiently. This represented a fundamental shift in the industry from being procedure-led to being much more client-led, and from being internally focused to being externally focused. As the management textbooks would describe it, the industry was moving rapidly towards a 'customer-focus' or 'market-led' approach.

The emergence of this type of expert private-sector client, who has the necessary in-house resources to manage large projects and a substantial ongoing construction programme, is one of the phenomena of the period, with the demand side of the industry despairing of the supply side ever putting its house in order. Clients had thus identified the need to develop bespoke methods of procurement, mainly based on the use of construction management techniques, in order to ensure that their needs were met effectively.

In this third phase, conventional methods remained the most widely used techniques, although there was a substantial increase in the use of design and build and a continued use of various forms of management approaches. There was some evidence, however, to suggest a reduction in the use of management contracting during the latter years of the period, which resulted from some clients' dissatisfaction with the performance of this method on large projects.

3.4 1990–2008

The fourth phase started around 1990, although the trigger for the recession that began to overtake the UK construction industry at about this time had its roots in the tightening of government monetary policy in 1988 and 1989 and the consequent withdrawal of property developers and industrialists from the capital development market.

Table 3.1 Trends in methods of procurement in the 1980s (by number of contracts)

	1985 %	1987 %	1989 %
Lump sum – firm BQ	42.8	35.6	39.7
Lump sum – spec. and drawings	47.1	55.4	49.7
Lump sum – design and build	3.6	3.6	5.2
Remeasurement – approx. BQ	2.7	1.9	2.9
Prime cost plus fixed fee	2.1	2.3	0.9
Management contracting	1.7	1.2	1.4
Construction management	–	–	0.2
Partnering agreements	–	–	–
TOTAL	100.0	100.0	100.0

Source: RICS 2010a.

The first half of the decade, in particular, was a time of low economic growth and uncertainty in business and finance, with social pressures mounting and environmental issues dampening enthusiasm for some major projects. Government capital spending, which peaked in the 1980s, was cut back and some 500,000 construction-related jobs were lost during the decade and more than 16,000 construction companies became insolvent, according to official figures.

Even in 1997, following the election of the New Labour government when there were some real signs of recovery and orders were at the highest level for over three years, the annual output of the industry was still some 20 per cent below the 1990 level, and it was not until the very last years of the decade that it could be said with any certainty that recovery had been achieved.

It was against this somewhat volatile background that the two defining construction industry reports of this post-modern era were produced. In 1994, *Constructing*

Table 3.2 Trends in methods of procurement in the 1980s (by value of contracts)

	1985 %	1987 %	1989 %
Lump sum – firm BQ	59.3	52.1	52.3
Lump sum – spec. and drawings	10.2	17.7	10.2
Lump sum – design and build	8.0	12.2	10.9
Remeasurement – approx. BQ	5.4	3.4	3.6
Prime cost plus fixed fee	2.7	5.2	1.1
Management contracting	14.4	9.4	15.0
Construction management	–	–	6.9
Partnering agreements	–	–	–
TOTAL	100.0	100.0	100.0

Source: RICS 2010a.

the Team was published, a government- and industry-sponsored report describing a wide-ranging review of the industry carried out by Sir Michael Latham, who was then a Conservative back-bench MP, although one with a keen interest in the construction industry (HMSO 1994a).

The report noted the dissatisfaction among clients with the service and products provided by the construction industry. In particular, it noted that projects were not delivered on time, ran over budget, were not of the quality expected and that too high a proportion of the project costs were spent on disputes and claims. The report emphasised the importance of the client's role, the need for better briefing of the design team, new and less adversarial forms of contracts, improved methods of selecting the team and more efficient ways of dealing with disputes. Altogether a further wake-up call.

The report's recommendations led to the development of the Construction Industry Board (CIB), which was set up in 1995, with the principal objective of implementing, monitoring and reviewing the recommendations of the Latham Report. Although no longer in existence, it had a significant impact on the industry in securing 'a culture of co-operation, teamwork and continuous improvement in the industry's performance'.

Four years later in 1998, the findings of the government's construction task force led by Sir John Egan was made public in the report *Rethinking Construction* (Department of Trade and Industry 1988). Both reports had considerable comment and analysis in the following years.

As stated above, the Latham Report was jointly commissioned by government and the industry, with the participation of many major clients, 'to consider procurement and contractual arrangements and the current roles, responsibilities and performance of the participants, including the client'. It was a report produced by a single individual, not a working party or committee; in the author's own words, 'It is a personal report of an independent, but friendly, observer.'

In essence, the review endeavoured to put forward solutions to problems it had identified that were preventing clients from obtaining the high-quality projects they required. The report concluded that an enhanced performance could be achieved only by teamwork in an atmosphere of fairness to all of the participants – the process of finding 'win–win' solutions.

Altogether, some thirty recommendations were made, ranging from the need for government to commit itself to becoming a best-practice client to the proposed introduction of a Construction Contracts Bill to give statutory backing to the recently amended Standard Forms of Contract, to ban many unfair practices, such as pay-when-paid clauses and to make dispute resolution procedures much quicker and less expensive.

Specific recommendations relating to the procurement process included:

1   The publication by government of a Contract Strategy Code of Practice, which, in terms that would enable it to be used by inexperienced as well as all other clients, should provide guidance on briefing, the formulation of a project strategy and procurement.
2   The production of a family of contract documents based upon the New Engineering Contract. The NEC (or Engineering and Construction Contract (ECC) as it was called in the second edition) was seen by Sir Michael Latham as a paradigm of

good practice because it was a project-management methodology rather than merely a set of contract terms and conditions.

3   The setting up of registers of consultants and contractors who have been approved to carry out certain types and values of projects.

4   Tenders should be obtained by public and private clients in accordance with European Union Directives and the National Joint Consultative Committee for Building (NJCCB) code of procedure. Design and build proposals should be obtained using only proposed procedures that are fair and reasonable to all involved parties. All bids should be judged upon the basis of the client obtaining best value for money, rather than just the lowest bid.

5   The use of partnering arrangements and framework agreements, subject to an initial competitive-tendering process, should be encouraged in the public sector.

Many other proposals, not directly related to procurement and covering training, dispute resolution, payment, design, teamwork on site, etc., were contained within the review, together with a programme of implementation involving both government and bodies representing the industry and clients. Although initially accepted in principle by all those involved, putting the numerous recommendations into practice was painfully slow. However, the Housing Grants, Construction and Regeneration Act 1996 (known as the Construction Act) was a direct consequence of the Latham Report and this has had a major impact on industry practices. Other aspects of the Latham Report were somewhat overtaken by the recommendations of the Egan Report, *Rethinking Construction*, in 1998.

Sir John Egan's report was commissioned by the UK government's Department of the Environment, Transport and the Regions (DETR) as a result of the growing dissatisfaction of both public and private clients with the performance of the industry, and the task force was asked to investigate and identify methods of improving the efficiency of the industry and the quality of its products.

The report was very critical of the industry for its poor past performance and called for a radical change in the way it implemented projects. The integration of the often separated processes of design and construction, increased standardisation, the use of lean construction, the cessation of competitive tendering and the eventual abolition of formal and adversarial building contracts were seen as the main means of achieving the necessary change.

Year-on-year targets of a 10 per cent reduction in construction costs, a similar percentage improvement in productivity and the same percentage increase in turnover and profit were proposed. A 20 per cent increase in the number of projects completed on time and within budget and the same percentage cut in defects and accidents were also called for.

Criticism of the report centred on its provocative and unnecessarily hostile approach and its failure to address the needs of occasional/one-off clients and the implementers of the small- to medium-sized projects which make up a large proportion of the industry's employers and therefore its annual workload. It was also noted that Sir John Egan's experience was predominantly in the motor car industry, where design and production are carried out by the same organisation and concepts such as standardisation of components and lean production are well developed. Critics also pointed out that construction does not take place in a controlled factory environment, and, indeed, the actual location of the 'assembly factory' (i.e. site) for each project necessarily changes.

In terms of the procurement process, the report's proposals advocated direct, trusting relationships being formed between clients, as the leaders of the project, and the other members of the team. It was implied that the need for more integration of design and construction might well lead to the breaking down of the existing adversarial barriers between the various parties involved in the project.

The increased use of partnering, framework alliances and similar co-operative procurement methods favoured by Latham appeared to be supported, although some commentators believed that the application of lean production theory, based on experience gained in the car-manufacturing industry, was central to the successful achievement of the task force's objective of increasing the construction industry's efficiency.

With government supporting and implementing its recommendations and participating in the multi-million-pound programme of demonstration projects that were started in 1999 by the Movement for Innovation (M4I – see section 3.4.1) in order to put the proposals to a practical test, it was hoped that the report's findings could be more effectively implemented than those of some of its predecessors.

Apart from these two major publications, a number of government reports were published during the decade, the majority of which dealt with procurement in construction either in a general or specific way. The first of these, *Partnering: Contracting without Conflict* (National Economic Development Office 1991) reported on a study of the practice of partnering based upon the experience of the construction industry in the USA.

The study concluded that the technique was beneficial to clients and to the industry as a whole and that there was ample scope for the continued development of the embryonic partnering activity currently taking place in the UK. A number of recommendations were made for the successful implementation of the method and a model form of contract was produced.

The following year, HM Treasury's Central Unit on Purchasing produced guidance on the selection of contract strategies (procurement systems) for major projects for the use of government departments (HM Treasury Central Unit on Purchasing 1992). The document, after reviewing and defining a contract strategy, described the way a project should be analysed, then explained and evaluated the various options for procuring the project, briefly outlined the selection process and provided a detailed plan for implementing the strategy.

While the terminology – particularly the use of the phrase 'contract strategy' – was somewhat confused, the setting out of a disciplined approach to the process of procurement system selection was to be applauded, especially when adopted by such a major client as central government.

In July 1995, the results of an efficiency scrutiny of the way in which government carried out construction procurement were published (Cabinet Office Efficiency Unit 1995). This examination had been commissioned by the Cabinet Office to establish the way in which government departments and agencies procured construction work and to recommend how it could ensure that government became a best-practice client.

This report should be seen as the vehicle for determining the most effective way of implementing the government's commitment given in a White Paper published two months earlier and as confirmation of its acceptance of the need to implement many of the recommendations of the Latham Report. The scrutiny looked at 20 major projects, examined 14 procurement systems used by government departments and consulted 200 people, nearly 50 per cent of whom were in the private sector, and concluded that the UK construction industry was 'in poor shape'. Specifically, the investigating team found the industry very adversarial, without customer focus, claims-conscious, internally fragmented and divided, undercapitalised with low profit margins and reluctant to adopt modern techniques for improving quality of service or the use of technology. So there had been precious little improvement since the first report in the early 1960s.

Government, as a major client, did not escape criticism, with all the projects that had been examined showing substantial increases over their approved costs and many failing to meet the planned completion date on time. There was a lack of understanding of the industry and insufficient involvement with the project or the consultants and contractors.

In an effort to rectify this state of affairs and to ensure that government became a best-practice client, 22 recommendations were made that laid down procedures for reorganising department procurement activities and methods, implementing intelligent risk management on all projects and increasing co-operation with the supply side of the industry to improve productivity and efficiency. All recommendations were accompanied by a deadline for their implementation.

Since 1995, there have been real attempts by government and the industry to tackle these shortcomings, and, in contrast to the lack of action that has been the fate of similar reports in past decades, efforts have been and are being made to achieve progress on the numerous proposals for improvement – only time will tell whether they will work.

In this fourth phase, the use of design and build increased much more rapidly than during the previous two decades, from under 10 per cent to over 20 per cent of the market share by number of contracts. Management-orientated methods fluctuated in their level of use, but generally increased their share of the market, with management contracting regaining ground so that, by value, it was back to its 1980s levels by the end of the century. The use of construction management also appears to be increasing compared with the immediate past and is clearly being used on the larger projects, as the market share in terms of value is higher than the market share by number of contracts. Partnerships and alliances appear to be on the increase, although definitive information is scarce owing to the comparative newness of the techniques used in the UK at the time.

Following the Egan Report in 1998, a lot of energy was put into improving the efficiency of the construction industry and almost resulted in an 'initiative overload' during the early years of the new millennium. Some of these initiatives are discussed below.

3.4.1 M4I

The Movement for Innovation (M4I) was established in 1998 to co-ordinate the implementation of the *Rethinking Construction* recommendations through the use of demonstration projects, working groups and knowledge exchange.

Demonstration projects were set up to benchmark project performance, be open and honest between the parties, share in the learning culture, set high standards in safety and respect for people and disseminate the results of their work to the rest of the industry through case histories, toolkits, and so forth. It had working groups covering: key performance indicators and benchmarking; the knowledge exchange; partnering the supply chain; culture change; education, training and research; sustainability and

Table 3.3 Trends in methods of procurement in the 1990s (by number of contracts)

	1991 %	1993 %	1995 %	1998 %
Lump sum – firm BQ	29.0	34.5	39.2	30.8
Lump sum – spec. and drawings	59.2	45.6	43.7	43.9
Lump sum – design and build	9.1	16.0	11.8	20.7
Remeasurement – approx. BQ	1.5	2.3	2.1	1.9
Prime cost plus fixed fee	0.2	0.3	0.7	0.3
Management contracting	0.8	0.9	1.2	1.5
Construction management	0.2	0.4	1.3	0.8
Partnering agreements	–	–	–	–
TOTAL	100.0	100.0	100.0	100.0

Source: RICS 2010a.

Table 3.4 Trends in methods of procurement in the 1990s (by value of contracts)

	1991 %	1993 %	1995 %	1998 %
Lump sum – firm BQ	48.3	41.6	43.7	28.4
Lump sum – spec. and drawings	7.0	8.3	12.2	10.0
Lump sum – design and build	14.8	35.7	30.1	41.4
Remeasurement – approx. BQ	2.5	4.1	2.4	1.7
Prime cost plus fixed fee	0.1	0.2	0.5	0.3
Management contracting	7.9	6.2	6.9	10.4
Construction management	19.4	3.9	4.2	7.7
Partnering agreements	–	–	–	–
TOTAL	100.0	100.0	100.0	100.0

Source: RICS 2010a.

respect for people. M4I, along with the Construction Best Practice Programme (CBPP) is now part of Constructing Excellence. See www.constructingexcellence.org.uk for more information.

3.4.2 KPIs

As mentioned earlier, the Egan Report set explicit and measurable targets for improvement. The year after the Egan Report, in 1999, M4I devised and published a set of key performance indicators (KPIs), which were subsequently refined and republished in April 2000. Eleven KPIs covered several aspects of the process of construction but, strangely, did not cover the performance of the product (i.e. the actual building). In April 2000 M4I piloted a set of sustainability indicators, which went some way towards redressing the balance between process and product. A wide variety of organisations have now developed these KPIs, most recently Constructing

Table 3.5 Trends in methods of procurement in the 2000s (by number of contracts)

	2001 %	2004 %
Lump sum – firm BQ	19.6	31.1
Lump sum – spec. and drawings	62.9	42.7
Lump sum – design and build	13.9	13.3
Target contracts	–	6.0
Remeasurement – approx. BQ	1.7	2.9
Prime cost plus fixed fee	0.2	0.2
Management contracting	0.6	0.2
Construction management	0.4	0.9
Partnering agreements	0.6	2.7
TOTAL	100.0	100.0

Source: RICS 2010a.

Table 3.6 Trends in methods of procurement in the 2000s (by value of contracts)

	2001 %	2004 %
Lump sum – firm BQ	20.3	23.6
Lump sum – spec. and drawings	20.2	10.7
Lump sum – design and build	42.7	43.2
Target contracts	–	11.6
Remeasurement – approx. BQ	2.8	2.5
Prime cost plus fixed fee	0.3	<0.1
Management contracting	2.3	0.8
Construction management	9.6	0.9
Partnering agreements	1.7	6.6
TOTAL	100.0	100.0

Source: RICS 2010a.

Excellence, which launched the 2006 Construction Industry Key Performance Indicators.

A report by the KPI Working Group for the Minister of Construction in January 2000 stated that clients of the construction industry want their projects delivered:

- on time
- on budget
- free from defects
- efficiently
- right first time
- safely
- by profitable companies.

It was also stated that regular clients are entitled to expect continuous improvement from their construction teams in order to achieve year-on-year reductions in both project costs and project times.

The purpose of the KPIs is to enable measurement of project and organisational performance throughout the construction industry, which can then be used for bench-marking and comparison purposes, and should therefore be a key component of any organisation's move towards achieving best practice.

Clients, for instance, will assess the suitability of potential contractors for a project by asking them to provide information about their performance against a range of indicators as part of the pre-qualification process. Some information will also be available through the industry's global benchmarking initiatives, so clients can see how potential contractors compare with the industry averages in a number of different areas.

While individual organisations have been measuring their performance for many years, there was too little consistency in the data and the way they had been published to allow comparison between organisations. Industry-standard KPIs were seen as a way of rectifying this deficiency and detailing a comprehensive framework for measurement.

Central government, the CIB and M4I continued to publish annual wall charts (see Table 3.7) for the headline KPIs and, where available, operational and diagnostic data were also published.

The KPI framework consists of the following main groups:

- time
- cost
- quality
- client satisfaction
- client changes
- business performance
- health and safety.

Within these groups, a range of indicators was developed to analyse either project or company performance, or both (see Table 3.7).

Once the contractor had measured its own performance in accordance with the above KPIs, the scores could be mapped onto the charts shown in Figure 3.1. Each chart for the individual KPI group gives a range and a median figure, showing whether the individual contractor is below average or above average in that area. It is clear how this procedure can help clients in their assessment of contractors for pre-qualification purposes.

Professional service firms were not exempt from this process, as there are now KPIs covering construction consultants in the areas of client satisfaction, productivity, profitability and training.

3.4.3 PFI

The private finance initiative (PFI) was introduced by the government in 1992 to enable major public projects to be funded without initially requiring the input of government capital funds. The purpose of the PFI is to deliver projects to the public

Table 3.7 List of key performance indicators issued by M4I in 2000

Group	Indicators
Time	1 Time for construction 2 Time predictability – design 3 Time predictability – construction 4 Time predictability – design and construction 5 Time predictability – construction (client change orders) 6 Time predictability – construction (project leader change orders) 7 Time to rectify defects
Cost	1 Cost for construction 2 Cost predictability – design 3 Cost predictability – construction 4 Cost predictability – design and construction 5 Cost predictability – construction (client change orders) 6 Cost predictability – construction (project leader change orders) 7 Cost of rectifying defects 8 Cost in use
Quality	1 Defects 2 Quality issues at availability for use 3 Quality issues at end of defects rectification period
Client satisfaction	1 Client satisfaction with the product 2 Client satisfaction with the service 3 Client's own criteria
Change orders	1 Change orders issued by client 2 Change orders issued by project manager
Business performance	1 Profitability of the company 2 Productivity of the company 3 Return on capital employed of the company 4 Return on value added of the company 5 Interest cover of the company 6 Return on investment for the client 7 Profit predictability of the project 8 Ratio of value added of the company 9 Repeat business of the company 10 Outstanding money on the project 11 Time taken to reach final account for the project
Health and safety	1 Reportable accidents including fatalities 2 Reportable accidents – non-fatal 3 Lost time accidents 4 Fatalities

sector, for example schools, prisons, hospitals and infrastructure works such as roads and bridges, together with the provision of associated operational services for a period of 25 to 30 years, after which they will revert to public ownership. The projects are funded and operated through a partnership of government and one or more private-sector companies.

There is a wide variety in the way PFI projects are organised, but they have some common features. Private consortia, usually involving large construction firms, facilities management firms and financial institutions such as banks are contracted to finance, design, build and manage new projects. The public-sector authority first signs

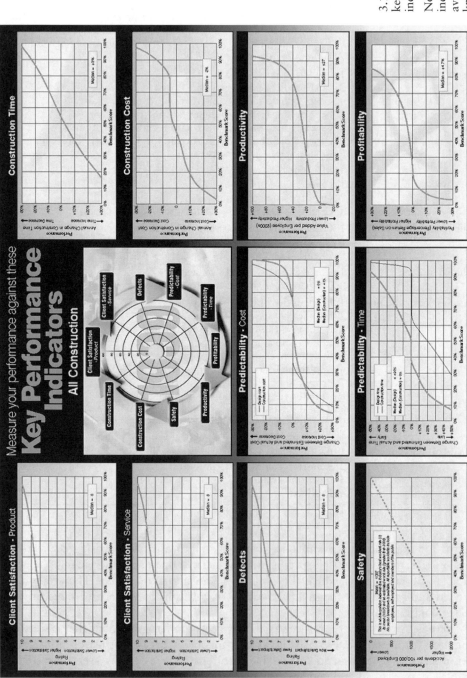

3.1 Construction key performance indicators (2000)

Note: updated indicators available at www.kpizone.com.

a contract with a private-sector 'Operator' who forms a special company called a 'Special Purpose Vehicle' (SPV) to build and maintain the asset. Chapter 11 gives more detail on the private financing of public projects.

3.4.4 Partnering and framework agreements

The concept of partnering had already emerged prior to 1994, but was given greater impetus by the Latham Report, which stated 'We are confident that partnering can bring significant benefits by improving quality and timeliness of completion whilst reducing costs.' A series of reports followed, including *Trusting the Team: The Best Practice Guide to Partnering in Construction* (Reading Construction Forum 1995); *Partnering in the Team* (Construction Industry Board 1997); and the RCF's follow-up report *The Seven Pillars of Partnering* (Reading Construction Forum 1998).

Partnering in the Team defined partnering as follows:

> *Partnering is a structured management approach to facilitate team working across contractual boundaries. Its fundamental components are formalised mutual objectives, agreed problem resolution methods, and an active search for continuous measurable improvements.*

It is clear from this definition that partnering was not intended as a particular type of contractual arrangement or procurement method, but rather as an approach to procurement, aimed at continuously improving performance over time, through team working, early identification and resolution of disputes and measurement of perform-ance against agreed objectives. Many organisations developed partnering charters and there are now several standard forms of partnering contracts, the most widely used of which are PPC 2000 (Project Partnering Contract) and its sister contract TPC 2005 (Term Partnering Contract), NEC3 Partnering Agreement (Option X12) and the PSPC (Public Sector Partnering Contract) issued by Be Collaborative, which is part of Constructing Excellence. The JCT has also issued a non-binding partnering charter.

Chapter 9 gives further details of partnering relationships.

Framework agreements may or may not be legally binding contracts themselves but an agreement with suppliers to establish terms governing contracts that may be awarded during the life of the agreement. The agreements therefore set out terms and conditions for future separate contracts (called the 'underlying contracts').

Chapter 10 gives further details of framework contracts, but we can note here that there are now at least four standard forms for framework contracts/agreements avail-able in the UK:

1   NEC3 Framework Contract (FC)
2   NEC3 Term Service Contract (TSC)
3   JCT05 Framework Agreement (FA)
4   NHS ProCure21 Framework Agreement (for the National Health Service in the UK).

3.4.5 Prime contracting

Prime contracting is a system launched in 2003 and used predominantly by government departments. The prime contractor is the single point of responsibility for an integrated process. There are generally two types of prime contract; the first is a contract for the operation and maintenance of all construction activity within a region (the Regional Prime Contract). The second type is a stand-alone contract for the design and construction of large projects which will include facilities management of the property and therefore will have a whole-life cost model, which must be complied with throughout a compliance period (the Stand Alone Capital Projects Prime Contract). The intention of this initiative is to encourage a more collaborative relationship between the client and the prime contractor and hopefully to deliver better value for the client.

3.5 Summary and tutorial questions

3.5.1 Summary

Like most industries which sell their goods to 'customers', the construction industry has been forced to change to being more customer-focused. Historically, the industry has been inward-looking and procedure-led, and these procedures were felt by their practitioners to guarantee a good-quality product, although little empirical evidence was available to support this notion, apart from the social and financial success of these practitioners. As clients became more professional, understood the construction process better and were forced to be more customer-orientated themselves, they naturally required this change of focus from their own suppliers and contractors. The move from almost total reliance on traditional procurement to a much wider variety of methods depending on clients' requirements has been quite a long and sometimes painful transition. This move has probably been helped by three economic recessions from the early 1970s, which taken together have necessitated a much leaner industry.

Allied to this change has been the movement towards performance measurement and KPIs. This movement is not restricted to the construction industry, as performance measurement and the 'tick-box' culture has pervaded all sectors and developed a whole new cadre of managers. Fortunately, this has now moved on from 'if it moves, measure it' to measuring and assessing only those issues which genuinely affect performance, however that is defined.

3.5.2 Tutorial questions

1   What do you understand to be the difference between open competitive tendering and selective tendering?
2   What are the possible dangers in a lack of co-ordination between design and construction?
3   Why would periods of recession in the economy in general have an impact on construction procurement methods?
4   What do you understand by the term 'adversarial contract'?
5   How could the move towards lean manufacturing help in changing construction procurement strategies?

4 Roles carried out in a construction project

There are various roles that must always be carried out in any construction project, irrespective of the size, location, form of contract or procurement route. The different procurement routes discussed in Chapters 6, 7 and 8 can be considered as merely different permutations of who actually carries out these roles for a particular project. For example, a building must be designed by someone and it must be constructed by someone. When the design is carried out by an independent architect and the construction by a building company, this is generally known as the traditional form of procurement (Chapter 6). However, when the building company takes responsibility for both the design and construction, this will be known as design and build (Chapter 8). The building company may decide to subcontract the design details to an independent firm of architects, but this is not the point – the building company is still *responsible* for the design.

Let us now look in more detail at the various roles in a construction project.

4.1 The client/client's representative

The construction client has been discussed at some length in both Chapter 1 and especially in Chapter 2. Clearly, without the client there would be no project and the client is therefore the main player in the project and one of the parties to the contract with the contractor. In most projects, their specific function is to:

- provide the site and allow possession of the site to the contractor during the period of construction;
- appoint professional advisers to carry out functions on their behalf that are essential for the design and administration of the project;
- give appropriate instructions to enable the work to be carried out;
- permit the contractor to carry out the work without interference;
- pay for the work properly carried out.

The client must not hinder the contractor in the progress of the work and must not interfere with any powers of the architect or contract administrator to issue certificates.

4.1.1 The employer's representative

As most clients are not knowledgeable about the construction industry and its processes (and why should they be, as they may well be industrial companies or retail organisations), then in order to administer the projects, they will appoint an agent or representative to help them or represent them.

This may be a person whose role is merely to attend meetings and inform the client of any matters that need to be brought to their attention. The client's representative does not normally have any powers themselves, depending on the specific form of contract entered into. It is interesting to note that under the design build contract, this role is known as the 'employer's agent' but the management forms of contract term it the 'client's representative'. Therefore we have an interesting semantic point regarding the difference between an 'agent' and a 'representative'. Under a normal definition, a representative can make decisions without constant recourse to the person appointing them (Members of Parliament are representatives of their constituents and therefore make the voting decisions themselves), whereas an agent has much more limited powers to merely negotiate on behalf of a client, with the final decisions being taken by the client themselves (e.g. an estate agent).

4.1.2 The employer's project manager

It is becoming common for employers to engage a project manager on more complex projects. This role may include managing both the design and construction stages and therefore cover both the design team and construction team. Some forms of contract, e.g. the New Engineering Contract (NEC) may give the project manager powers to give instructions, take decisions, etc.

4.2 Designer

The designer, who will normally be an architect, or an engineer for structural designs, is usually responsible for designing the works (see Chapter 8 on design and build, where the design may have been started by an architect but taken over by the contractor, under a procedure called 'novation'). While it is usually the case that the designers go on to carry out the duties and obligations of construction-stage supervision, this is not necessary in all cases, and the client is free to appoint different firms for this responsibility.

The principal duties and responsibilities of the designer normally include:

- issuing copies of contract documents;
- determining any levels and other essential setting-out data;
- issuing further information and instructions necessary to enable the contractor to carry out the works;
- explaining ambiguities;
- issuing variation instructions where necessary;
- requiring the removal of work that is not in accordance with the contract;
- issuing instructions regarding the expenditure of provisional sums;
- issuing instructions regarding the making good of defects that appear within the defects correction period;
- valuing work done for payment purposes and issuing payment certificates;
- issuing certificates of substantial (or practical) completion of the works or sectional completion;
- issuing certificates of making good defects;

- issuing the final certificate;
- granting extensions of time.

In addition to the above duties and obligations, which are normally written into the contract, the architect, or engineer, will usually be responsible for the following:

- dealing with statutory undertakings, i.e. electricity, water and other utility providers;
- arranging and chairing project team meetings;
- monitoring progress and reporting to the employer;
- co-ordinating any design carried out by other consultants or the contractor and any subcontractors.

4.2.1 *The clerk of works*

A clerk of works is an inspector of the works usually employed by the designers or by the supervision consultants if this is a separate appointment. They can issue directions to the contractor but have limited powers and must usually be confirmed by the architect or contract administrator before they have full contractual effect. The clerk of works is generally responsible for seeing that the work is carried out in accordance with the contract specifications and good practice.

Outside the UK, separate testing companies can be appointed to provide a technical supervision service, which may encompass and extend the clerk of works role. For example, in France, an organisation called a *bureau d'étude* is often appointed for the detailed supervision of construction and will also carry out technical tests such as concrete cube testing, structural stability calculations and M&E design checks.

4.3 Design Manager

The Design Manager is a relatively new role within the construction team, especially in the UK, as the architect or other lead designer has invariably taken on the role of managing the design process. The key areas of responsibility for the design manager are to:

- manage the performance of the various design consultants to maintain the contract scope, cost, schedule and the objectives of the project;
- challenge the design consultants to maximise quality and value while minimising the risk profile for the project and the overall programme;
- manage the co-ordination of interfaces with the work of the various design consultants;
- manage third-party and client technical and constructability reviews of deliverables;
- manage the client and other stakeholder sign-offs where appropriate during the design stage;
- provide assistance and advice in the preparation of project requirements and objectives, and co-ordinate these across the programme;
- manage designer procurement, contracts and performance, ensuring that the designer's responsibilities and scopes of service are clearly set out and terms and conditions of contract are consistent with client guidelines;

- ensure that design deliverables are defined for each stage of work;
- ensure that a clear and realistic design programme is in place;
- ensure that all designers are aware of, trained in and consistently use project processes and procedures (e.g. design standards and procedures and design systems and tools);
- ensure that all designers are capable of delivering solutions consistent with client sustainability and health and safety programme goals and objectives;
- ensure that cost of design services are commensurate with required design effort and industry norms;
- ensure that performance indicators are in place to measure designers' performance with respect to the above;
- manage the design consultant's contract with respect to design delivery;
- administer the following key project processes and manage the consultant for compliance:

  - risk management
  - change management
  - project review and assurance
  - document control
  - health and safety in design (CDM)
  - confidentiality
  - value-management/value-engineering workshops
  - engage the technical resources required to challenge the designers to obtain the most beneficial balance of quality, value and risk for the overall programme
  - identify potential interfaces with other projects and facilitate co-ordination of design and construction issues among the appropriate parties
  - manage third-party and client technical and constructability reviews of deliverables
  - engage and lead the technical resources required to provide quality assurance on the consultant's deliverables, covering aspects such as completeness, cost, buildability, risk, health and safety, sustainability and specialist technical issues
  - contribute to the client's procedures for ensuring 'fitness for purpose'
  - manage the consultant to obtain detailed planning permission; third-party and statutory approvals
  - manage the response to consultant's RFIs
  - provide regular progress reports on all of the above to the project manager.

4.4 Design cost control

The purpose of design stage cost control is primarily to ensure that the final design solution that is accepted by the client can be constructed within the client's original budget. The way this is done is through a process known as 'cost planning' and is generally carried out by the consultant quantity surveyors employed as part of the design team.

The RICS *New Rules of Measurement* 2010b gives the formal cost estimating and cost planning stages (see Table 4.1), using the 2007 RIBA *Outline Plan of Work*.

Table 4.1 The RICS formal cost estimating and cost planning stages in context with the RIBA *Plan of Work* and OGC Gateways (adapted from RIBA 2007)

		RIBA Work Stages	RICS formal cost estimating and elemental cost planning stages	OGC Gateways (applicable to building projects)	
Preparation	A	Appraisal	Order of cost estimate	1	Business justification
	B	Design brief		2	Delivery strategy
Design	C	Concept	Formal cost plan 1	3A	Design brief and concept approval
	D	Design development	Formal cost plan 2		
	E	Technical design	Formal cost plan 3	3B	Detailed design approval
Pre-construction	F	Production information	Pre-tender estimate		
	G	Tender documentation			
	H	Tender action	Post-tender estimate		
Construction	J	Mobilisation		3C	Investment decision
	K	Construction to practical completion		4	Readiness for service
Use	L	Post-practical completion			Operations review and benefits realisation

4.5 Constructor (contractor)

The constructor's role is clearly to actually build the project, in accordance with the design and specification produced by the designers and to the cost and time that they have negotiated and agreed in the tender and contract documents. Generally, the constructor is referred to as the contractor in most projects and is the party engaged by the client (employer) to construct the works (and if required, to design all or part of the works). Once selected and engaged to execute the work described in the contract, the prime obligations of the contractor include:

- carrying out and completing the works in accordance with the contract;
- the contractor may also be required to design part of the project, which is cleverly known as the 'contractor-designed portion'.

In addition to the above primary obligations, the terms of the contract will impose other obligations or amplify these primary obligations. The contractor is also normally required to:

- take care of the works and, if required, to take out insurances to cover any indemnities given by the contractor;
- comply with authorised instructions given under the contract;
- comply with the drawings and specification for the works;
- proceed regularly and diligently with the works;

- complete the works within the period stated or within a reasonable time if no time is stated (this is a nightmare scenario and can cause havoc with contract administration, therefore projects should always have a specific time period for completion – known as the contract period – which is not the same as the contractor's programme);
- give notices required by the contract with regard to delay or additional payment.

It is normal for the contractor to engage subcontractors and suppliers to execute part of the work and/or to supply goods and materials. Let us also take a look at their function:

4.5.1 Subcontractors and suppliers

As we have already seen, as recently as the 1970s, contractors executed a considerable amount of work themselves, subcontracting only that work that they were not geared up to do or that they could have done cheaper by others. Common trades that were subcontracted included plastering, painting, glazing, roofing, plumbing, heating, electrical and other specialist work. Today, it is common to find brickwork, joinery, excavation, concrete work and almost all other trades being subcontracted. The modern main contractor has become more or less a manager of other subcontractors, or work packages as they are also known.

Domestic subcontractors

Domestic subcontractors are usually selected entirely by the main contractor. While most forms of contracts require the architect's or engineer's approval, this is usually a formality, and the contractor is given complete freedom to select their own subcontractors if no specific requirement to subcontract is given.

The function of domestic subcontractors is to carry out that part of the works that is the subject of the subcontract in accordance with the directions of the contractor in such a manner as to ensure that the contractor does not commit a breach of the main contract.

Named subcontractors

In most cases, named subcontractors are treated as domestic subcontractors. The contractor may be given a limited choice (a number of firms may be listed in the contract from which the contractor may select a domestic subcontractor) or they may have no option but to use a single firm named in the contract.

The use of named subcontractors enables the designer to have some control over the selection of firms to carry out selected work.

Nominated subcontractors (not appropriate for JCT05)

As the term implies, these subcontractors are selected and nominated by the employer or their agents. Apart from rules regarding valuation and payment for work done by nominated subcontractors and (in the case of JCT98 contracts) consent to grant extensions of time, the nominated subcontractors are very little different from domestic subcontractors. There is a tendency for contractors to consider that nominated subcontractors are 'nothing to do with them' (probably as a result of the false sense of security where extensions of time may be granted in the event of default by the subcontractor).

The other major area of the main contractor's 'supply chain' is that of the suppliers of materials and plant. Both contractors and subcontractors rely on a wide range of suppliers for the goods and materials to be incorporated in the works and, again, these are categorised as 'domestic' and 'named/nominated' suppliers.

Domestic suppliers

Unless the suppliers are nominated or specified, the contractor or subcontractor will be free to purchase materials from any supplier they choose provided that they comply with the specification. Approval of suppliers may be a condition of the contract, but this will usually be a formality unless there are special provisions regarding source or sample, etc.

The supplier's function is to provide the specified goods in accordance with the contractor's requirements (delivery dates, quantities and price, etc.). Contracts of supply will usually contain provisions regarding ownership and title of the goods as well as replacement of defective goods, etc.

Named or specified suppliers

Named suppliers are not regarded in the same way as named subcontractors. Therefore, if the employer wishes to name a supplier, they can usually only do so by specifying the supplier in the specification, in which case the contractor is bound to use that supplier for the specified goods or materials. There may be good reasons to specify a supplier, but unless it is absolutely necessary, this practice should be avoided. Once the supplier knows that they are the only one in the race, the price tends to go in one direction (and that's not down). Additionally, the practice may be contrary to competition and fair trade or anti-trust legislation.

Nominated suppliers

As the name implies, nominated suppliers are also selected by the employer to supply certain goods and materials. Under the ICE conditions of contract, for example, there is no distinction between nominated subcontractors and suppliers, and the contractor may make reasonable objection to either. The rules relating to nominated suppliers in JCT98 (i.e. clause 36) have been deleted in their entirety from the 2005 edition of the contract.

4.6 Construction management/supervision

In the traditional procurement system, where there is a main contractor, they are responsible for the construction management of the project during the execution stage when on site. Care must be taken, however, when the project is procured through one of the other separated routes to ensure that all site management roles have been covered.

The Construction Management Association of America (CMAA) says the 120 most common responsibilities of a construction manager fall into the following seven categories:

1   project management planning
2   cost management
3   time management
4   quality management
5   contract administration
6   safety management
7   construction management professional practice, which includes specific activities such as defining the responsibilities and management structure of the project management team, organising and leading by implementing project controls, defining roles and responsibilities and developing communication protocols, and identifying elements of project design and construction likely to give rise to disputes and claims.

However, in the UK and other national systems developed from it, the construction manager would be responsible for only the following items in the above list:

1   project management planning
2   time management
3   quality management
4   safety management
5   construction management professional practice.

Cost management and contract administration are covered by quantity surveyors or commercial managers within the construction organisation.

The client will also often require a specific consultant role to supervise the construction management of the various contractors, which will cover all of the above seven areas in the CMAA list.

4.7 Cost control during construction

Cost control is the activity that compares actual cost or expenditure against planned costs, adjusting one or the other dynamically by reference to the project's financial environment. It is a process that should be continued throughout the construction period to ensure that the cost of the building is kept within the approved cost limits set by the agreed contract sum plus any necessary additions issued by the client. This process can also be known as earned value analysis (EVA).

In many large projects, professional construction cost consultants or quantity surveyors play a critical role in estimating construction costs, scheduling construction activities and implementing various techniques to complete the projects. It is very easy for the cost of construction to exceed the approved budget, one of the major problems that need to be addressed by clients, construction management consultants and contractors – i.e. to minimise cost overruns, the client, project managers and contractors need to improve their skills and abilities in dealing with project cost control. During economic recessions, there is added pressure on construction companies for better and more effective cost control so that any waste (or other non-value added costs) is reduced to as close to zero as possible.

There are generally three components that cause project cost overruns:

1   incomplete specifications and drawings
2   requirements of the owner
3   changes from contractors.

Although project management teams try to co-ordinate all requirements and deal with the above issues, there are several factors in all projects that make overruns almost inevitable and certain factors that are both beyond their ability and outside their responsibility.

The common causes of cost problems are:

- poor estimating techniques and/or standards, resulting in unrealistic budgets;
- out-of-sequence starting and completion of activities and events;
- inadequate work breakdown structure;
- no management policy on reporting and control practices;
- poor work definition at the lower levels of the organisation;
- management reducing budgets or bids to be competitive;
- inadequate formal planning that results in unnoticed, or often uncontrolled, increases in scope of effort;
- poor comparison of actual and planned costs;
- comparison of actual and planned costs at the wrong level of management;
- unforeseen technical problems;
- schedule delays that require overtime or idle time costing;
- unforeseen material cost inflation.

Therefore during the construction stage, the client or the client's representative must ensure that the project management team, designers and/or consultants obtain their approval before issuing any variation or change orders that have cost significance. As mentioned earlier, the client should also establish project objectives and project expectations clearly and carefully from the beginning of the project and these should be consistent through each project stage.

For effective cost control during the construction stage, the factors can be split into three groups.

Project management

The client must co-ordinate with the project management team on the following five major project issues: (1) programming and functional needs of the facility; (2) site questions and concerns; (3) building systems requirements; (4) construction schedule; and (5) project budget.

The owner and the project management team have to ensure that each issue is reconciled with the others and updated frequently. If any issue is ill defined, there is a greater chance that the project will go out of control.

Project change control

When changes must be made that will change the plan, the project manager must control and track them. The guidelines for controlling changes are as follows:

1   Changes can be requested by anyone associated with the project, but submitting an approved Project Change Request Form to the project manager and the project change committee will ensure that only authorised people can approve changes.
2   Project change requests must be evaluated to determine the cost effect of a request.
3   The project manager and the team must estimate the effort, dependencies and resources required to institute the project change before it is formally approved.
4   The project manager must communicate the impact of proposed changes in terms of cost, scheduling and performance to the client and all parties who are to implement the change.
5   If a change is approved, the signatures of the client and the project manager are required authorising the project team to make the requested change.
6   The project plan must be revised after the change has been approved for implementation.
7   The project manager must inform the project team of all approved, unapproved and deferred changes.
8   The Project Change Request Form must be logged within a rigorous document control system, regardless of its approval or rejection.

Cost control

In a complex project, effective cost control should be approached as an application of Pareto's law, which essentially states that 80 per cent of the outcome of a project is determined by only 20 per cent of the included elements. Thus in establishing a cost-control system, the idea is to isolate and control in detail those elements with the greatest potential impact on final cost, with only summary-level control on the remaining elements.

Cost-control problems come about for many reasons, including incorrect estimating techniques, predetermined or fixed budgets with no flexibility, schedule overruns, inadequate work breakdown structure (WBS) and so forth. Good project management planning techniques during the planning processes will prevent cost problems later in the project.

Cost control, like many of the other controlling processes, is concerned with monitoring project performance for variances or differences. The client and their advisers should always try to keep control of the factors that cause change and variance in order to keep their impacts to a minimum. Cost control ensures that all appropriate parties agree to any changes to the cost baseline. This process is ongoing in order to manage cost changes throughout the project.

For the contractors on a project, it is important to have a formal cost-control system for the project, as many will be paid on a lump-sum basis and in order for them to maintain their required profit margin, it is only the costs that can be effectively controlled internally.

4.8 Handover and acceptance

Once the construction has been completed to the client's satisfaction, there is a stage of handing over the facility to the client, or whoever is going to use it. Therefore completion and handover are very much interlinked. This is the final stage of work by the contractors and the consultants before the owner or tenant takes responsibility for the facility. Normally, the completion and handover are carried out under the continued supervision of the project management team, but this may not necessarily be so. In some countries, such as France, a separate set of consultants is employed to approve that the completed facility is in accordance with the design. These consultants have had no input to the design of the building or the contractual relationships and therefore can inspect the facility more objectively.

The following actions are required at this stage:

- Ensuring that the contract administrator has inspected the works and issued the certificate of completion (sometimes known as practical completion, substantial completion or preliminary handover) with the list of outstanding snags/defects and a timescale for their final completion. These snags should not be serious enough to interfere with the client's use of the facility, as they should be only of a minor nature. In some countries, a snagging list is also known as a punchlist
- Advising the client that the issue of the certificate of completion marks an end of the contractor's responsibilities to secure and insure the site. Furthermore, the client cannot issue instructions or variations after this date.
- The final account can now be concluded with the contractor as they are deemed to have finished the works – except for the snags.
- The defects liability period (sometimes known as the defects correction period or defects notification period) commences from the issue of the certificate of completion, and normally lasts twelve months but could be longer for more complex projects. Within this period, the client can require the contractor to fix any defects that manifest themselves, and at the end of the period, the final certificate is issued to the contractor after a final inspection. After the final certificate is issued, the contractor is absolved from all responsibilities under the contract. They are free to go.

These stages will be replicated if the project has sectional completions (partial possessions), i.e. if the client takes over parts of the facility at different times as a phased handover.

At the date of practical completion, a number of significant documents are handed over from the contractor to the client:

- the project's health and safety file
- 'as-built' drawings together with all relevant specifications of the finished works
- operating and maintenance manuals for any installed plant and equipment
- warranties and guarantees from suppliers
- copies of statutory authority approvals and consents
- testing and commissioning documents.

4.9 Health and safety (CDM) roles

Legislation concerning construction health and safety in the UK is largely covered in the Construction (Design and Management) Regulations. These, together with other health and safety legislation, have completely changed the nature of construction management in the UK since 1995. A glance through the contents pages of any of the standard forms of contract will also show how they have impacted on contract administration and management by the references to the planning supervisor (renamed 'co-ordinator' in the 2007 Regulations), principal contractor, health and safety plan and health and safety file.

The Construction (Design and Management) Regulations 1994 (CDM Regulations) brought new responsibilities to all those involved in the construction process. They built upon and clarified existing duties under the Health and Safety at Work, etc. Act 1974 and the Management of Health and Safety at Work Regulations 1992. The CDM Regulations outline a series of duties that make clear how all parties can contribute to improving the management of health and safety on a construction project. The assurance of proper competence and resources lies at the heart of the roles and responsibilities of all the parties. Clients, designers, supervisors, principal contractors and other contractors will have to meet their responsibilities and be able to prove that they have done so. This will require documentation of the procedures that all parties have implemented.

4.9.1 The CDM 'client'

The client has responsibility for applying the Regulations at each key stage of a project, defined in the 1994 Regulations as:

- deciding whether the Regulations apply;
- appointing a competent planning supervisor with adequate resources;
- providing the CDM co-ordinator with relevant information about the *premises* where construction work is to be carried out;
- appointing a competent principal contractor with adequate resources;
- approval of the health and safety plan before sanctioning construction work to commence;
- taking custody of the health and safety file and making it available for inspection.

These duties have been augmented by additional duties under the 2007 Regulations:

- *Management arrangements.* All clients must make appropriate arrangements to ensure that the construction work can be carried out safely and that fixed workplaces will be safe to use. This includes the provision of welfare facilities for all site staff, whether employed by the main contractor or any of the subcontractors/work-package contractors.
- *Appointments.* Under CDM 2007, the client must appoint a CDM co-ordinator before significant detailed design work begins, i.e. before the initial concept design (stage C of the RIBA *Plan of Work*; see Table 6.2 in Chapter 6). Crucially, CDM 2007 states that if the client does not appoint the CDM co-ordinator and also principal contractor at the appropriate time, they will be deemed to have appointed themselves in these roles, which should concentrate their minds somewhat!

- *Information*. Clients must provide both designers and contractors with any details or information regarding health and safety in order to identify hazards and risks associated with the works (this is called the preconstruction information). On larger 'notifiable' projects, this information is given to the CDM co-ordinator, who will ensure that it is passed on to the appropriate party. There is a new client duty in CDM 2007 regarding the minimum mobilisation period allowed to the contractors, which must also be included in the F10 notice to the HSE, where the project is notifiable.
- *Commencement*. As with the 1994 Regulations, the client must ensure that construction does not start until a compliant construction phase plan has been prepared by the principal contractor, which must set out the contractor's organisation and arrangements to manage risk.
- *Health and safety file*. The client has a new duty to provide information for the health and safety file to the CDM co-ordinator and to ensure that the information can be easily identified, kept available for use and is revised regularly.

4.9.2 The CDM 'designers'

When the client has decided to go ahead with a project, it is the responsibility of the designer to ensure that the client is aware of their duties under the Regulations. Additionally, on notifiable projects, the designers must ensure that a CDM co-ordinator has been appointed before they start work on the substantive design.

There are *three* principal duties under Regulations for designers:

1   to notify the client of their obligations under the Regulations;
2   to reduce the health and safety risks to anyone carrying out construction work or cleaning/maintaining the works. The wording of the 1994 Regulations has been clarified in the 2007 Regulations to reduce any possible ambiguities;
3   to co-operate with the CDM co-ordinator and any other designers.

4.9.3 The CDM co-ordinator

The Regulations impose a duty on clients to make two statutory appointments on construction projects. The first of these is the CDM co-ordinator who is appointed by the client to provide management input from a health and safety point of view. As mentioned above, the co-ordinator must be appointed before any substantive design work is carried out on the project. They take no direct responsibility for health and safety in either the design or construction phases of a project and are required to possess only two credentials, competence and adequate resources.

The CDM co-ordinator therefore co-ordinates health and safety in construction and has the following duties prescribed by the legislation:

- to ensure that the HSE is notified of a project if applicable;
- to review design work as it progresses for compliance with the Regulations;
- to ensure that designers co-operate with each other;
- to provide 'suitable and sufficient' advice to clients and contractors with respect to the Regulations;
- to ensure that a health and safety plan and file are prepared and that they are adequate.

It must be emphasised that the CDM co-ordinator is not responsible for:

- preparing designs
- preparing health and safety documentation
- monitoring health and safety on site

although there might be a need in some circumstances for the co-ordinator to be proactive on health and safety issues on the site.

The extent of the CDM co-ordinator's duties is therefore potentially very wide, and the advice and assistance required from them will vary from project to project depending on the client's expertise and existing management arrangements, together with the scope of the project. What is 'suitable and sufficient' on one project may be totally inadequate on another.

4.9.4 The principal contractor

The principal contractor may or may not be the main contractor. In certain procurement routes, it may be one of the work packages, or even one of the consultants. Whoever is stated in the contract documents as the principal contractor, their duties are to:

- plan, manage and monitor the construction phase in liaison with other contractors;
- prepare, develop and implement a written plan and site rules (the initial plan would have been completed before the construction phase begins);
- give the various contractors the relevant parts of the plan;
- make sure suitable welfare facilities are provided from the start and maintained in good order throughout the construction phase;
- check the competence of all appointees;
- ensure all workers have site inductions and any further information and training needed for their work;
- consult with the workers regarding any risks or safety issues;
- liaise with the CDM co-ordinator regarding any ongoing design;
- secure the site.

4.9.5 Other contractors

Other contractors will be engaged on the construction site, as subcontractors or work-package contractors. They also have particular responsibilities and duties on all sites, which are to:

- plan, manage and monitor their own work and that of their workers;
- check the competence of all their appointees and workers;
- train their own employees;
- provide information to their workers;
- comply with the specific requirements in Part 4 of the Regulations;
- ensure that there are adequate welfare facilities available for their workers.

Furthermore on larger notifiable projects they must:

- check that the client is aware of their duties and a CDM co-ordinator has been appointed and HSE notified before starting work;
- co-operate with the principal contractor in planning and managing work, including reasonable directions and site rules;
- provide details to the principal contractor of any contractor whom he engages in connection with carrying out the work;
- provide any information needed for the health and safety file;
- inform the principal contractor of problems with the plan;
- inform the principal contractor of reportable accidents, diseases and dangerous occurrences.

In addition, all workers on site are required to check their own competence to do the work, co-operate with others and co-ordinate work so as to ensure the health and safety of construction workers and others who may be affected by the work and to report any obvious risks.

4.10 Summary and tutorial questions

4.10.1 Summary

There are various roles that must be carried out in any construction project, and procurement routes can be most easily classified according to who carries out these roles. Table 4.2 shows these roles and who would normally carry them out in the procurement routes described in this book.

Apart from the employer, all of these roles represent the 'supply side' of the industry, i.e. those companies supplying the product required by the purchaser (client). As each construction project is unique, each one therefore has to be designed individually.

Table 4.2 Roles in a construction project in various procurement routes

	Separated roles (Chapter 6)	Overlapping roles (Chapter 7)	Integrated roles (Chapter 8)
Client/client's representative	Employer	Employer	Employer
Designer	Architect	Architect	Contractor
Design manager	Architect	Client's rep./PM	Contractor
Design cost control	PQS	PQS	Contractor
Constructor	Contractor	Contractor	Contractor
Construction management	Contractor	Contractor/PM	Contractor
Construction supervision	Architect	Client's rep./PM	Client's rep./PM
Construction cost control	PQS/contractor	PQS/PM	Contractor
Handover and acceptance	Architect	PM	Client's rep./PM
Health and safety	CDM co-ordinator and principal contractor		

When the design is complete, the construction is then planned and carried out. In some cases, the construction of a section of works can be carried out when the design of that section is complete irrespective of whether the full project design is complete.

Although each project is unique, the processes by which the design, planning and construction are carried out are similar for every project, so the working relationships between the parties can be developed and refined over time. Therefore each player in the supply chain more or less knows what it has to do, what risks it has accepted and what all the other parties do, together with their risks. Complications naturally arise when these definitions change or different conditions of contract are introduced that allocate the risk in slightly different ways. It is essential therefore that in any given project, the parties are aware of the responsibilities of all concerned, and these responsibilities should be clear and unambiguous.

4.10.2 Tutorial questions

1   Why is the client (employer) the most important party to a construction project?
2   What are the major differences between an 'order of cost estimate' at the beginning and a 'pre-tender estimate' at the end of the cost planning stage?
3   Consider the difficulties associated with a subcontractor being 'nominated' or 'named' by a party who will have no direct contractual relationship with them.
4   What are the advantages of separating the carrying out of the works (construction) with the management of that construction?
5   Is cost control during the construction stage more effective than cost control during the design stage?

5 Tendering and payment

There are three major methods of payment for construction work, or any other type of services for that matter. First, each party can agree a price for the work before it starts and that price will be a set figure (known as a lump sum). Unless there are any variations, this is what the builder (contractor) will be paid, i.e. it is a fixed price. Clearly, in this case, the design should be sufficiently progressed for the tendering contractors to be able to accurately estimate the cost of the works, add their required profit margin and submit a fixed-price tender. Second, the design may not be sufficiently progressed, but the designers know what items are required and can ask the tendering contractors to submit rates for these items and they will be paid the actual amounts when work is completed at the rate for these items. This is known as a firm-price contract, as the rates are firm but the total contract value cannot yet be fixed. It is also known as a 'remeasurement' contract as the drawings are remeasured after completion of the works (although, paradoxically, they may not have been measured in the first place). Third, the design may be totally vague or the speed at which the work is carried out may be crucial and the client wishes to appoint the contractor early so will appoint them on a cost-reimbursable basis whereby all the contractor's project costs are repaid together with a percentage addition to cover overheads and profit.

Let us now look at these three variants in more detail.

5.1 Lump-sum (fixed-price) contracts

As mentioned above, this type of contract requires the design to be at or near completion, and tender documents are sent out to contractors who wish to submit a bid (tender). The tender list can either be 'open', in which case any firm can submit a tender but may have to pay a deposit for the tender documents, returnable on submission of a bona fide tender. Alternatively, the client may have chosen a 'selective' list of tenderers after going through a pre-qualification procedure to assess their competence to do the work and whether they have sufficient resources for the size of project envisaged.

Lump-sum contracts are therefore normally carried out in a single stage and from a selected list of tenderers who have been pre-qualified to ensure that they have the sufficient resources and capabilities. Contracts of this type and procedure are thus also known as 'single-stage selective tendering'.

The tender documents required for a lump-sum fixed-price tender are:

a   tender drawings
b   detailed specifications of material and workmanship required on the project
c   form of tender
d   form of contract and amendments, if any
e   tender instructions.

See Chapter 6 for further discussion of this topic.

The price for which the tendering contractor is willing to do the work is given in the form of tender as a single figure (the lump sum). For example, JCT Practice Note 6 gives a Model Form of Tender in Appendix C where the tendering contractors insert a lump sum and contract period and the whole form is only two pages in length.

This is all very well for relatively small jobs where the tendering contractors will be able to estimate their costs from the supplied drawings and specification. However, for larger projects, it is normal to provide some form of pricing document so that the client or their advisers can see how the lump sum has been calculated and consequently carry out a tender analysis. This pricing document can also be used during the construction phase for payment purposes.

For 'plan and specification' projects, therefore, the tender documents will comprise those listed above, but larger projects will also include a pricing document such as a bill of quantities, schedule of rates or activity schedule – see section 5.4 below.

The published rules for tendering of lump-sum fixed-price contracts are very strict, as it is important to be as fair as possible to all tenderers and these rules have been set out in the various Codes of Procedure for Tendering, published by the NJCC (National Joint Consultative Committee for Building, which sounds very formal and bureaucratic, and it was). The NJCC as an organisation was disbanded in 1996, although its publications are still in use today. For JCT contracts in the UK, the rules for main contract tendering are covered in Practice Note 6 published by JCT Ltd.

No client is legally obliged to use these codes of procedure and can choose a contractor on any criteria they wish. However, the codes represent good practice, and the client is more likely to obtain a competitive tender by operating in accordance with them.

5.1.1 The tender list

After it has been decided that a contractor is to be selected by competition for a lump-sum contract, a shortlist of suitable tenderers will be drawn up either from the client's own list of approved firms or from the responses to an advertisement in the media. A quick glance through the technical press such as *Building* magazine, *Construction News* or *Contract Journal* will show advertisements for projects, or inclusion on a client's approved list. The EU requires larger projects to be advertised across Europe in the *Official Journal of the European Union* (*OJEU*) and, not surprisingly, this adds considerable time to the entire process.

Estimating and tendering for construction projects is an expensive business for construction firms and the larger the tender list, the greater will be the abortive costs of tendering by those firms who are not successful. Somebody has to pay these costs, which are generally added to the overheads of projects that the firm does win, thereby increasing tender levels overall. Additionally, large tender lists mean that each firm has a reduced chance of winning (one in ten when there are ten tenderers against one in three when there are only three tenderers). Therefore the NJCC Code of Procedure recommends that competitive tender lists should contain no more than six firms. Consideration should also be given to the amount of work demanded of the tenderers during the tender period. For design and build projects that require much more contractor input at tender stage, therefore, the maximum number of tenderers should be four.

As mentioned above, it is common practice now for experienced clients to develop a list of approved contractors by requiring them to go through a pre-qualification process, which will assess the following:

a   The firm's financial standing and record over the last three years or so.
b   Whether the firm has had recent experience of building at the required rate of completion over a comparable contract period.
c   The firm's general experience, skill and reputation in the area in question.
d   Whether the technical and management structure of the firm including the management of subcontractors is adequate for the type of work envisaged.
e   The firm's competence and resources in respect of statutory health and safety requirements.
f   The firm's approach to quality assurance systems.
g   Whether the firm will have adequate capacity at the relevant time.

Approved lists should be reviewed periodically to take out firms who may no longer exist, whose performance has proved to be unsatisfactory and also to allow the introduction of new firms and personnel. It is always good practice with any database to clean it regularly.

5.1.2 Tendering procedure

Preliminary enquiry and tender documents

To enable contractors to decide whether they wish to submit a tender and to anticipate demands on their estimating department, each potential tenderer should be sent a preliminary invitation to tender and the codes of tendering procedure include standard templates for these invitations. When the contractor has confirmed their agreement to tender, the tender documents will be forwarded to each of the tendering contractors. The conditions of tendering must be absolutely clear so that all tenders are submitted on the same basis and can therefore be easily compared against each other. See above for a list of the documents that should be included in the tender pack (tender documents).

Regarding the actual conditions of contract, it is not unusual for clients to amend the standard forms to suit their own circumstances, and in international contracts it is most unusual for clients not to amend the standard conditions. Tendering contractors are used to this and know from experience that when a client amends a standard form of contract, they do not amend it in favour of the contractor. Therefore this adds an extra risk to the contractor and the tender often reflects this extra risk. The standard forms of contract, whether they are JCT, ICE, NEC or FIDIC, have been carefully considered and carefully worded by experts from across all sides of the industry, so why is there any need to amend them? An interesting question for clients' lawyers.

Time for tendering and tender compliance

The time allowed for the preparation of tenders should be determined in relation to the size and complexity of the project. Inadequate tendering time often leads to mistakes – if the mistake increases the tender price, the tenderer may not win the job, and if it reduces the tender price, the tenderer may well win the job but not make the profit they anticipate. A period of four weeks is normally the minimum, although for major projects the period should be longer. The latest time for submission should be clearly stated in the tender documents and should specify an hour and a day with any tenders received after that time being rejected by default.

If bills of quantities or other pricing documents are included in the tender documents, it is not normal for them to be required when submitting the lump-sum tender. If a contractor's tender is to be considered further, the bills of quantities will be called in for analysis.

If a tenderer considers that any of the tender documents are deficient, ambiguous or require further clarification, they should inform the party issuing the documents as soon as possible, who will then send the appropriate clarification to all tenderers. Of course, this is all very well in theory, but most contractors deliberately look for the ambiguities and adjust their tenders accordingly. For instance, if the drawings show considerably more quantity in the project than is included in the bills of quantities (clearly a mistake by the quantity surveyor), they are likely to insert a higher rate than normal in the bills, which will have a moderate effect on the tender, but when the mistake is seen during the construction stage and a variation order issued to increase the quantities in the bills, the contractor will make a healthy profit on this item. The contractor is required to build from the drawings and specification, as the drawings normally have a higher status than the bills of quantities as a contract document.

Under English law, a tender may be withdrawn at any time before it is accepted, but this is not necessarily the case in other countries.

For fair competitive tendering, tenders submitted by each tenderer must be based on identical tender documents and tenderers should not attempt to vary the basis of tendering by qualifying their tender.

Qualified tenders

Under most legal jurisdictions, a contract must contain an offer and an acceptance of that offer. It is possible to invite another party to make an offer (called an invitation to treat) and under English law, a contract must also contain something of value given by both sides (called 'consideration'), although again this is not necessarily the case in countries where there is a civil code. Therefore in construction tendering:

a   The client invites several contractors to tender for a project.
b   The tenderers offer to build the project for a sum of money (offer with consideration).
c   The client accepts one of those offers after analysing all tenders.

Under (a) above, the invitation is sent out with the tender documents attached. It is therefore easy to see why the offer by the tenderers must be compliant with the tender documents, otherwise it is a qualified tender and amounts to a counter offer. If the client receives qualified tenders or alternative offers that have varied any aspect of the project specification or contract period, these should be rejected or the client faces possible legal action by the other tenderers for unfair advantage.

Assessing tenders and notifying results

It goes without saying that tenders should be opened as soon as possible after the time for receipt of tenders, as the tender process is very expensive for contractors and they need to know the results reasonably quickly. Normally, the lowest two or three tenders are required to submit their pricing documents, so that they can be analysed for arithmetical errors or other more technical issues. Those tenders outside the lowest three should be notified that they have been unsuccessful and given an indication of where their tender figure came, so they can make improvements to their estimating and tendering procedures. In the UK, it is now illegal to put in a 'cover price' for a tender as this is considered to be collusion. A cover price is an artificially high tender price put in with the intention of not winning the job. This was not uncommon practice by contractors who did not wish to carry out a particular project (possibly because they did not have the resources at the time) but did not wish to tell the client for fear of being removed from their approved list. The UK government's Office of Fair Trading considers that cover pricing is a form of cartel where a group of contractors would agree between themselves who will win which projects. In late 2009, several major contractors received multi-million pound fines for this practice.

In the UK, under health and safety legislation, a contract must not be entered into or work started on site until the contractor's competence and resources have been satisfactorily assessed.

Examination of the pricing document

The examination of the priced bills of quantities or activity schedules should be carried out by the client's consultant quantity surveyor (PQS). These documents are obviously confidential as they include pricing and cost information of the tenderer. The purpose of examining the pricing documents is to detect any arithmetical errors in building up to the tender figure. If there are any errors there is generally a set procedure to allow the contractor to either stand by their tender figure or adjust it accordingly, with the consequent risk of either losing the project or losing profit.

The PQS will also analyse the tender in a more technical way by looking at the spread of rates throughout the project. If the tendering contractor has 'front-loaded' their tender, by which most of their profit is included in the early trades, such as groundworks or structural frame leaving the latter trades, such as decorating, to be carried out at cost or below, then suspicions may be aroused regarding the contractor's intentions and possibly financial stability. If the contractor cannot finish the project for whatever reason, the client will be unlikely to find another contractor willing to use the same rates for the latter trades.

Negotiated reduction of tender figure

A contractor's tender and its build-up should never be altered unilaterally by the client or their consultants, especially when no modifications have been made to the scope of works or specification since the tender documents were issued. However, should the lowest tender, after the tender analysis, still exceed the client's budget, the recommended procedure is for a reduced price to be negotiated with the tenderer based on changes to the scope and specification. All negotiations should be fully documented.

A further way of reducing the tender figure that gained some notoriety in the mid-decade around 2005 was the practice of 'reverse auctioning'. After the client has received all tender figures, they will announce the lowest figure to all tenderers, giving them a short period of time to confirm a lower price. If the tenderers accept this challenge and reduce their bid, the process may well start again, hence the term 'auction' and as the price goes down (rather than up, as in a normal auction), it is referred to as 'reverse auctioning'. Many in the industry see this as an unethical development, and the more reputable and professional contractors have refused to engage with the practice.

The 'credit crunch' recession appears to have encouraged the development of this practice, and there are now internet-based systems used by clients that allow tenderers to revise their bids (downwards) after submitting their initial tender. Consider the following article taken from the *Building* magazine website.

Contractors attack rise of 'eBay' tendering

26 February 2010
By Sophie Griffiths

Fears have been raised by contractors over the resurgence of procurement methods such as 'eBay tendering' that are intended to arrive at the lowest possible price.

This method, which is based on the auction website eBay, works by inviting companies to submit bids online and compare them with those of their competitors. The bids can then be revised downwards before a cut-off date.

Many have expressed concerns that clients increasingly favour such procurement routes as the recession forces them to cut costs.

This process, which is also called a 'Dutch auction', is common in bidding for the supply of commodities, but the rise of their use for services in the recession has angered many.

Paul Jessop, chief executive of the Federation of Plastering and Drywall Contractors, said: 'Our members raised this at our annual meeting earlier this month. This is a dangerous route for clients to take. People will end up putting in stupid bids because of the pressure of the auction.'

Stephen Ratcliffe, director of the UK Contractors Group, agreed that the industry needed to be wary of such methods, which he said had pitfalls in terms of health and safety, an argument supported by pressure group Families Against Corporate Killers, which has also criticised the trend.

Bill Taylor, managing director of East Midlands Plastering, said he would avoid social housing contractor Keepmoat after it told him it would use this method more, and base 90 per cent of each decision on price. He said: 'We're not a commodity but a service. With this we're only as good as the next price.'

A spokesperson for Keepmoat said: 'The Keepmoat E-Procurement system is transparent and provides best value procurement solutions to our customers. The process includes the evaluation of both cost and qualitative criteria to give a balanced selection. The response from our supply chain has been very positive and we have listened to their feedback when developing the system. We believe our system to be best practice, and in collaboration with our supply chain we aim to develop it even further.'

Supermarket Asda is understood to be among the clients looking at eBay tendering, while Tesco has reportedly used it in the past.

Michael Tiplady, international director at Jones Lang LaSalle, said another new trend was three-round tendering, where bidders are told whether they offered the lowest price at the end of each round and given the chance to revise it. He said such methods left no margin for manoeuvre. 'If extras are needed, firms will be forced to go to the client for money.'

(Source: http://www.building.co.uk/story.asp?storycode=3158724&origin=bldg
weeklynewsletter.)

Letters of intent

Letters of intent are very dangerous and should be avoided wherever possible unless they are written very precisely.

Letters of intent are used when the client wants to accept a tender but cannot actually sign the contract yet. They may be waiting for a statutory approval or the contract can only be signed by the chairman or CEO but nevertheless want to go ahead with the works or ordering materials with a long delivery period. The original purpose of a letter of intent was merely to inform the contractor that their tender was successful and that a contract would be entered into in due course. A carefully worded letter of intent would not form a binding contract, i.e. it does not amount to acceptance of the contractor's tender.

Recently, the original purpose of a letter of intent has changed to enable the employer to enter into an agreement with the contractor on limited terms (for example, authorising the contractor to carry out some design work, to order materials and fabricate structures) that do not amount to acceptance of the tender but that keep alive the employer's option to accept or reject the tender at a later date. Care should be taken when drafting a letter of intent, and contractors who receive a letter of intent should take equal care in deciding whether to accept its terms. If the contractor decides to proceed on the basis of a letter of intent, it is essential that they are in no doubt as to the limit of authority given by it and how payment will be made.

The primary legal case in English law regarding letters of intent is *British Steel* v. *Cleveland Bridge*, where the court had to decide whether a letter of intent created a contract. The judge had this to say:

> *Now the question whether in a case such as the present any contract has come into existence must depend on a true construction of the relevant communications which have passed between the parties and the effect (if any) of their action pursuant to those communications. There can be no hard and fast answer to the question whether a letter of intent will give rise to a binding agreement; everything must depend on the circumstances of the particular case.*

It was decided that it did not matter whether a contract came into existence. If one party acted on the instructions given in a letter of intent and was simply claiming payment, then they were entitled to be paid on a *quantum meruit* basis (meaning 'a reasonable price for the work').

Such a state of affairs is unsatisfactory from both parties' point of view. One party may be under the impression that the work will be paid for at rates stated in the documents that were intended to be incorporated in a contract at some future date, while the other party may believe that payment will be on the basis of actual cost plus reasonable overheads and profit. Neither view may be correct and the courts may have to decide what is reasonable.

So, what should a letter of intent say? The most important aspects to be addressed are:

- precise instructions as to what work is to proceed and the specification required;
- terms of payment for the work to be done;
- provision for termination of the work covered by the letter of intent or for the employer to exercise his option to accept the contractor's tender;
- the employer's rights to the benefit of any orders placed pursuant to the letter of intent in the event of termination;
- ownership of materials ordered pursuant to the letter of intent;
- liability for loss or damage; insurance.

It is easy, therefore, to see why letters of intent are dangerous unless written by experts.

5.1.3 Guaranteed maximum price (GMP) contracts

A GMP contract is exactly what it says on the tin, the client and contractor establish a price for a specific scope of work that cannot be exceeded. GMP contracts are generally used on fast-track projects or when the design is incomplete at the time construction starts meaning that a fixed price based on tender drawings, specifications and bills cannot be calculated with any degree of accuracy.

The GMP project team consists of the client, the architect, the consulting engineers, quantity surveyor, other specialist consultants and the contractor. Emphasis is placed on teamwork and the contractor is invited to participate in design-team meetings from an early stage of the project as they are taking the risk of any cost overruns.

The contractor provides the owner with a guaranteed maximum price to manage the construction, which includes all their prime costs (labour, materials, plant and subcontractors), plus their own overhead and profit. The contractor provides regular estimates during the design process to evaluate the building costs and ensure that the GMP is achievable with the design team using this information to progress the design.

The main advantage of GMP is to reduce the overall design and construction schedule to meet a deadline. Construction is usually started in phases to allow drawings to be completed progressively, allowing the programme schedule to be reworked to manage issues that arise during design or construction.

Value-engineering workshops can also be used to identify design alternatives to help the project maintain budget and schedule. Workshops should be held at key design-phase milestones to allow alternatives to be evaluated and incorporated into the design. As with other collaborative procurement routes, all project-team members are encouraged to participate in the value-engineering workshops.

The scope of work should be adequately defined for pricing at the end of design development. The contractor then prepares an overall estimate of the project that should be lower than the GMP, as this price cannot be exceeded. The contractor is responsible for any cost overruns but may be required to share any savings with the client. The contractor will also normally include a contingency to allow for refinement of the design during the construction stage but not for new scope, which will be subject to additional negotiation.

5.1.4 Target cost contracts

Target cost contracts in some ways occupy the same intermediate position as GMP between lump-sum and reimbursable contracts, as the financial risk is shared between the client and contractor. Under a target cost contract, the actual cost of completing the project is compared with a target cost previously agreed. If the actual cost exceeds the target cost, some of the cost overrun will be borne by the contractor (known as the 'painshare') and the remainder by the client in accordance with an agreed formula. Conversely, if the actual cost is lower than the target cost, then the contractor will share the saving with the client again in accordance with a previously agreed formula (known as the 'gainshare'). Such an approach helps to align the interests of the parties since both will have an interest in working together in order to reduce the costs of the project to an achievable optimum.

Despite this, claims under target cost contracts can be among the most difficult to manage, which is due mainly to the nature of the contracts themselves. Currently, there is only one published standard form of target cost contract in use in the construction industry, NEC3 Options C and D. This form of contract has, however, not secured universal acceptance and is often heavily modified, and the implications of the target cost mechanism are not fully appreciated by those who prepare contracts so that there is doubt as to the effect of certain situations on the target cost. When there is a significant sum at stake, the effect of this doubt may well be to create a dispute.

The areas giving rise to the greatest difficulty in practice are, first, the target cost itself. The formula for determining painshare or gainshare must be recorded precisely in the contract documents. A mathematical formula is the best way of achieving this, as it is clearly more precise and objective. Problems do, however, occur in circumstances in which an aggressive painshare formula is imposed on the contractor, and in such circumstances the cost risk on the contractor can be almost as great as under a lump-sum contract with a subsequent effect on the likelihood of poor working relationships and contractor's claims.

Agreement on the target cost should only be reached when the client's requirements have been defined to a sufficient degree to enable a target cost to be ascertained with some accuracy. The client's fundamental requirements for the project will need to be fully described in the project brief, all necessary information made available and all risks identified. A target cost agreed in advance of this stage is less likely to give the right mix of reward and incentive to the contractor. It may well be that the client's requirements have been sufficiently developed at the tender stage to allow the target cost to be agreed, even though they are still short of the level of detail needed for a full lump-sum contract. In many cases, however, the client's needs can only be ascertained following extensive discussions with, and design work by, the contractor. A mechanism therefore needs to be included in the contract for paying the contractor for their initial pre-target cost work. There may need to be a bonus element in the formula for payment as many of the opportunities for cost saving occur at this very early stage.

Additionally, it is important to have clarity regarding the categories of cost that are, and are not, to be included in the definition of actual cost and target cost respectively. Consideration may need to be given in particular to the treatment of prime-cost sums (if used), contingencies, free-issue supplies from the client (and their delivery dates) and the contractor's head office overheads and profit.

As with all construction projects, it is inevitable that variations will be made after the target cost has been agreed. These will invariably have an effect on the cost of the work, so the contract needs to make provision for the target cost to be adjusted accordingly. Where a variation originated by the client increases cost, this is normally acceptable to both sides. There is a difficulty, however, with those variations that result in a reduction in cost. If the target cost is reduced by the amount of the reduction, the whole of the benefit of the reduction in cost is obtained by the client. This can be regarded by the contractor as unjust, particularly if it is the contractor who has suggested the variation as part of their value-engineering workshops. A mechanism therefore needs to be found to reward the contractor in these circumstances.

When target cost arrangements are initially agreed, they should be broken down in sufficient detail to enable the amount of the cost increase or reduction to be ascertained without difficulty. A target cost recorded as a single figure may be insufficient for this purpose and therefore could lead to disputes. It is also important that changes in target cost are agreed promptly. Uncertainty as to the position will inevitably reduce the incentive for the contractor, and a dispute is far more likely to occur if an attempt is made to agree an adjustment long after the event.

Closely related to variations is the procedure for the approval by the project manager of documentation or designs produced by the contractor. Conflict often arises if the project manager is perceived as using this process to impose their own design preferences on the contractor, with additional cost being incurred as a result, rather than using the variation procedure, which would allow the target cost to be amended. While this approach by the project manager may be accepted by the contractor under a reimbursable contract, since the client will bear the additional cost, things will be viewed very differently under a target-cost regime. The contractor in these circumstances will have to share in the additional cost, and any requirements of the project manager are therefore likely to be scrutinised by the contractor with considerable care.

Any legislation that comes into effect after the agreement of the target cost and affects the actual cost figure (e.g. changes in taxation) will also need to be provided for, and a decision will be required as to whether the additional cost should be borne entirely by one of the parties or shared between them through the target price mechanism.

Finally, the contract may on occasions be terminated by the client before completion. That in itself may well give rise to a dispute, but the likelihood is even greater if there is no clear statement as to how the contractor's painshare/gainshare is to be dealt with in this situation. To deprive the contractor of any gainshare in these circumstances would be clearly unfair, as it may encourage the client to terminate shortly before completion and hence avoid payment to the contractor of a substantial gainshare. It would therefore be preferable to make some kind of assessment of painshare/gainshare in relation to that part of the project that has been completed.

All of this may suggest that the preparation of a target cost contract is a somewhat daunting task. Certainly it can prove demanding for those who are not aware of all the implications of this form of contract and therefore using a standard form of contract written for the purpose is strongly advised.

5.2 Firm-price contracts

As mentioned earlier, a firm-price contract refers to the rates in the pricing document (bill of quantities or activity schedule) that are 'firmed up' at the contract negotiation stage. The quantities may not necessarily be settled, so the total lump sum cannot be agreed and it is not therefore a fixed-price contract (as fixed price refers to the total value of the project). So, having clarified that point, what are the different firm-price contracts available to the client?

A firm-price contract would use either a bill of approximate quantities or schedule of rates as the pricing document (see sections 5.4.2 and 5.4.3 below for a fuller description of these documents).

For these types of contract in the UK, the standard form is the JCT Standard Building Contract with Approximate Quantities (SBC/AQ) issued as revision 2 in 2009 and (in its own words) is appropriate:

- *for larger works designed and/or detailed by or on behalf of the Employer, where detailed contract provisions are necessary and the Employer is to provide the Contractor with drawings; and with approximate bills of quantities to define the quantity and quality of the work, which are to be subject to remeasurement, as there is insufficient time to prepare the detailed drawings necessary for accurate bills of quantities to be produced; and*
- *where an Architect/Contract Administrator and Quantity Surveyor are to administer the conditions.*

Can be used:
- *where the Contractor is to design discrete part(s) of the works (Contractor's Designed Portion);*
- *where the works are to be carried out in sections.*

As the pricing document contains only approximate quantities, in order to establish the actual cost of the works, 'remeasurement' is needed, as stated at the beginning of this chapter. This means that when the work is actually carried out, the contractor and PQS will agree the actual measurements of the finished work and substitute those quantities for the approximate quantities in the bill. At the end of the project, the bill of quantities will therefore be 'as built'.

As the design was not fully developed at the time of tender, extra items may need to be included in the bill of quantities, and, conversely, items in the bill may not be needed. In the latter case, the items would be merely deleted (providing it did not affect the contractor's profit margin too much) but in the former case, a new rate would have to be agreed between the parties that should be calculated on a similar basis to the remaining rates in the bill, or if this is not possible by agreeing a fair rate. The use of dayworks to calculate the cost of an item should be avoided if at all possible, as this effectively converts the item to a cost reimbursement (or cost-plus, as it is also called). As mentioned elsewhere, dayworks should be used only as a last resort in pricing construction work.

5.3 Cost-reimbursable contracts

Cost-reimbursable contracts are increasingly being used in construction, which has become necessary because the conventional lump-sum, fixed-price contracts are not always sufficiently flexible to deal with the diverse demands of clients and contractors. However, there are two major weaknesses of simple cost-reimbursable contracts: first, the lack of knowledge of their overall financial commitment by the client; and second, the lack of incentive for the contractor to control their costs. The lack of knowledge of overall financial commitment is clearly related to the lack of definition of the client's requirements at tender stage and is generally independent of the contract type.

However, cost-reimbursable contracts also have many advantages for both the client and contractor. First, design and construction can progress simultaneously, which can lead to early completion and thereby reduce the inflation effect on capital cost and to some extent interest charges on borrowings. A further advantage in the case of a revenue-earning facility is the benefit of receiving early income from rentals or sales. To achieve these benefits, an effective and practical cost and schedule control system must be established for the project in order to deliver the project on time and within budget. Such a system should provide the information required by the project team to compare the actual progress with the planned progress and use the techniques of earned value analysis (EVA) or variance analysis.

Consequently, the project team can verify whether work is in line with or deviates from the original plan. This also highlights contractor performance and any labour and plant productivity, indications that are essential for keeping control of the project and, if necessary, for identifying corrective measures.

Therefore, in cost-reimbursable contracts the contractor is paid their actual costs (sometimes called prime costs) including preliminaries, together with a fee to cover their overheads and profit, which may be either a percentage fee or fixed fee. This payment mechanism is appropriate where an early start is required but the project lacks sufficient definition to allow a fixed price or firm rates to be established. Cost reimbursement has also been used in projects where the particular physical conditions are considered too variable to allow normal methods of payment to be adopted, and the overriding consideration has been to ensure the full and open co-operation of the contractor to allow the construction problems to be overcome. Indeed, cost-reimbursement contracts have been adopted where the client perceives that they have all the necessary construction expertise to make the major decisions and only the contractor's resourcing skills are needed.

Cost-reimbursable contracts create shared risks between the client and contractor and therefore have a major effect on the relationship between the parties. The main part of the financial risk is clearly with the client, who has to fund all actual costs of the project and has no accurate indication of the final out-turn costs. This means that the contractor may have little incentive to work efficiently and economically, unless of course they are interested in further work from this client. It is therefore in the interests of the client to ensure that the contractor is encouraged to co-operate in forecasting the final out-turn costs, so that joint action may be taken at the appropriate time to prevent any cost overrun.

There are two main means of achieving this. The first is to create a legal relationship that requires the contractor to notify the client when they have reason to believe there will be a cost overrun. This is the approach adopted in the USA, where doctrines of good faith and fair dealing have developed that go beyond those in the UK through anti-trust legislation. The second approach is to share the risk of cost between the client and contractor by using target cost contracts or maximum guaranteed price. The former is the most common form of cost-reimbursement contract in UK construction.

5.4 Pricing documents

5.4.1 Bills of quantities

Bills of quantities (BOQs) have existed in one form or another for over 300 years, and even today, debate over the relative advantages and disadvantages of BOQs is long-standing and generates strongly held and often conflicting views. The BOQ is a document that itemises the finished work in a construction project. It is usually prepared by a consultant quantity surveyor employed by the client (often known as the PQS), based on detailed drawings and specifications prepared by the project architect. The BOQ has two primary uses:

- In the pre-contract stage, the BOQ assists contractors in the preparation of their tenders. The BOQ breaks down the contract works in a formal, detailed, structured manner that the tendering contractors can use to build up their estimate of the cost of the works.
- In the post-contract or construction stage, the BOQ assists both the contractor and PQS in the valuing of progress payments and variations. The BOQ provides a financial structure for contract administration.

A BOQ can be prepared using various alternative standard methods of measurement depending on the nature of the project and its complexity. In the UK, the main standard methods of measurement are:

- The Standard Method of Measurement of Building Work, 7th edition (SMM7)
- The Civil Engineering Standard Method of Measurement, 3rd edition (CESMM3)

There are also other specific methods of measurement for highway and bridge works and some larger clients have also developed their own methods usually based on one of the above main methods.

Internationally, many countries have developed their own standard methods of measurement, e.g. Ireland, Australia and Malaysia, which are more appropriate for the particular regulatory conditions in that country as well as for different construction techniques.

The contractual status of the BOQ can vary in that it can form part of the contract documents with the quantities being considered firm or it could be provided as an approximate bill of quantities that require remeasurement during the construction period. In the latter case, the approximate BOQ would not form part of the contract documents.

Historically, the PQS's workload has been predominantly reliant on the production of BOQs as well as post-contract work of interim valuations, pricing of variations and settlement of final accounts, with tender documentation accounting for a considerable proportion of their workload. However, over recent years, there has been a significant decline in the PQS's workload associated with producing BOQs as a result of the increasing use of non-traditional forms of procurement that do not use formal BOQs, the reduction in fees obtained for preparing the documents and the relative ease of outsourcing their preparation to areas of the world with reduced unit costs.

The production of a full BOQ 'taken off' from a fully detailed design requires considerable time to prepare. Many clients are reluctant to give this time or do not understand that they need to allow the design team adequate time to prepare a detailed design and the subsequent documentation for tendering. In particular, the amount of additional time to prepare a full BOQ can be offset by a reduction in tendering time, particularly on larger projects, and would also lead to more competitive and accurate tenders with fewer ambiguities and therefore less opportunity for disputes during the construction stage of the project. The process of producing a BOQ, however, requires the PQS to interrogate the design and specification in considerable detail, which also enables them to identify inaccuracies and inconsistencies in the drawings and specification prior to tender, also in turn helping to further reduce any subsequent post-contract problems.

As mentioned above, the BOQ provides a common basis for both the production and comparison of tenders. The structured format simplifies the analysis of each tender build-up, and even when they are not provided by the client's consultants each tenderer often prepares their own quantities, so the measurement effort is multiplied by the number of tenderers. All contractors will need to know the extent and quantity of work in a project, so some measurement must of necessity take place. The absence of a BOQ may lead to greater variability, increased risk in estimating and consequently more disputes during the construction stage or encourage the contractor to cut corners in an attempt to recover the consequent loss.

Bills of quantities therefore generally have the effect of reducing the costs of tendering by up to 5 per cent, depending on the size of the project. Main contractors and subcontractors like having BOQs in the tender packs and consider that firm BOQs increase the competitiveness of their tenders while 'plan and spec.' tenders can increase tender prices because of the increased risk.

A greater number of subcontractors are likely to submit tenders for works packages when there are BOQs as part of the tender pack. This is mainly due to the fact that most subcontractors are relatively naïve about the commercial aspects of construction work, and the clearer structure of BOQs helps them to price the work competitively, especially if they normally use rates from previous projects for tendering purposes.

Advantages and disadvantages of bills of quantities

The main advantages of bills of quantities are:

1   *Simplified tender analysis.* All tenders can be analysed on the same basis and each tenderer's rates can be compared to fair rates in the marketplace at the time.
2   *Calculation of interim valuations and progress payments.* The rates are used to value the work completed to date during the construction stage and make progress payments to the contractor.
3   *Valuing of variations/change orders.* The rates in the BOQ are used for valuing variations and changes that have been authorised by the project manager, whether as additions or deductions. Therefore the variations are valued on the same basis as the original tender.
4   *Assessing the final account.* The final cost of the work will be based on the rates in the BOQ.
5   Database. The pricing details within the BOQ provides a cost database for future feasibility estimating and cost planning.
6   *Fee calculation.* The BOQ provides an absolute basis for the calculation of consultants' fees, if the fees are based on percentage of construction costs.
7   *Asset management.* The BOQ provided readily available data for asset management of the completed building, life-cycle costing studies, maintenance schedules, general insurance and insurance-replacement costs.
8   *Taxation.* BOQs provide a basis for quick and accurate preparation of depreciation schedules as part of a complete asset management plan for the project.

The disadvantages, on the other hand, include:

1   *Cost and time.* The preparation of a BOQ tends to increase the cost and lengthen the design period or documentation period.
2   *Estimating practice.* Tenderers may ignore the formal specification document by pricing only according to the BOQ. This may lead to under-pricing and the consequent risk of unsatisfactory performance. The specification document is part of the design and therefore normally has a higher status than the BOQ.
3   *Procurement.* The use of a detailed design and associated BOQ may discourage contractors from submitting alternative design solutions, as alternatives will amend the quantities. A firm BOQ is suitable only to the traditional procurement system.
4   *BOQ errors.* Because the BOQ is a complex document and developed from a design that may not be 100 per cent complete, there are likely to be errors, omissions and discrepancies between the drawings, specification and BOQ. The contract should make it clear which document has priority.

Structure of bills of quantities

Bills of quantities are normally structured in accordance with the standard method of measurement used for the project. In the UK, when using SMM7, the structure will generally be in the following sections:

1   preliminaries
2   prime cost and provisional sums
3   preambles
4   measured work.

Preliminaries

These are the general items usually associated with the contractor's site establishment on the project. A look through section A of SMM7 will show various general cost items included in the project. As preliminaries are effectively site-based overheads, their costs are not related directly to the quantity of work but rather the duration of the project and the method adopted to construct the works. Preliminary items fall under the following headings:

* general employer's requirements
* limitations on method, sequencing or timing
* temporary works and services
* contractor's management and staff
* site accommodation
* contractor's mechanical plant
* any works or materials supplied by employer.

The contractor will normally price the preliminary items as a lump sum, but in some cases they are required to price the item as a time charge or an event charge. For example, the cost of tower cranes will be (a) erecting the crane, (b) rental for the period it is on site and (c) dismantling the crane; none of these costs relates to the amount of work done but to either an event or a time period.

Prime cost and provisional sums

Prime cost sums (PC sums) are a procedure to include the cost of a nominated subcontractor or a nominated supplier into the main contractor's tender figure and contract sum. They are rarely used now in the UK, as modern standard forms of contract do not recognise the concept of nomination following the judgments in various legal cases. A nominated firm was a firm that the client or architect effectively instructed the main contractor to appoint as a subcontractor or supplier. Their costs, which acted as a prime cost to the contractor (i.e. the equivalent of labour, plant and materials), would be covered by the PC sum and the main contractor was entitled to add a percentage profit to this sum and also allow for general or special attendances on the subcontractor (i.e. provision of welfare facilities, health and safety responsibilities, power supplies, etc.). When the nominated subcontractor's final account was received, this replaced the PC sum in the main contractor's final account. Although the nominated firm was a subcontractor, the main contractor had limited power over them.

Provisional sums were originally sums included in the BOQ for work that had not yet been fully designed but for which an allowance is required in the contract sum. Contingencies are included in project costs as a provisional sum. Provisional sums have now taken over the role of the PC sum, in that organisations that would have been covered by a PC sum are now covered by a provisional sum, such as statutory undertakings – those organisations who are the only ones allowed to do certain work, such as connecting to the mains electricity, mains gas, etc.

As the main contractor is also required to programme the works, having an amount of work in the project that is described as provisional means that they cannot fully programme all the works if they do not know its total extent. SMM7 therefore separates provisional sums into two categories. 'Defined provisional sums' mean that the contractor has included the scope of the work as part of their programme and therefore cannot claim an extension of time or loss and expense as a result of the architect or contract administrator firming up the actual scope, even if it increases the cost of the sum. 'Undefined provisional sums', on the other hand, mean that the contractor has not included the extent of the works in their programme and therefore may be able to claim for additional time or cost. Not surprisingly, many clients insert a clause in their contracts to the effect that all provisional sums are defined.

In section 5.2, the concept of dayworks was mentioned. Dayworks will be included as a provisional sum and is intended to be used to value work where no other method is appropriate, i.e. as a last resort. Dayworks rates will be included for labour, plant and materials with the contractor inserting their all-in rates (i.e. including all statutory on-costs) and then adding a percentage addition to cover the disruption of taking labour from their planned activities to work on the daywork instruction. As these percentage additions can be anything up to 150 per cent, many contractors are unsurprisingly quite keen to price work as dayworks and can get quite upset when this is refused even after their timesheets have been signed by the clerk of works.

Measured work

The main body of a bill of quantities contains measurements of the amounts of finished quantities of materials in a project. These quantities are normally structured in sections in accordance with the 'Common Arrangement of Work Sections', part of the Co-ordinated Project Information family of documents, which is now incorporated into the Uniclass framework (Unified Classification for the Construction Industry). These sections roughly relate to modern trades, so that the main contractor may separate out the sections for distribution to specialist domestic subcontractors.

In some cases, the BOQ will be structured in elemental format rather than trade format as described above. Elements refer to the parts of the building rather than the trade specialisations, with Table 5.1 giving the list of both elements and work sections taken from the relevant section of Uniclass.

The preparation of BOQs has changed considerably over the past thirty years, mainly as a result of the ubiquitous use of computers for both design and tender document production. Prior to the 1970s, the measurements were 'taken off' the drawings by the quantity surveyor and set down on 'dimension paper'. The calculations of volumes, areas, lengths and weights known as 'squaring' was carried out by a comptometer operator (remember that hand-held calculators were not widely available until the mid-1970s and would have been useless anyway prior to metrication of the industry in 1968, since all measurements were then in feet and inches). The items and their quantities were put into bill order by another specialist quantity surveyor called an 'abstractor'. This all took some considerable time, but the system had so many self-checks that mistakes were rare and could usually be traced back to design inconsistencies.

From the mid-1970s, the laborious task of abstracting all but disappeared as the new system of 'cut and shuffle' gained ground. In this system, all the individual measurements were written down on smaller dimension sheets, which were then shuffled into bill order at the end of the taking-off stage. This invariably meant that there were thousands of pieces of paper spread around the office as they were put into trade or elemental order, and one gust of wind from the window could wreck a day's work. The actual typing of the BOQ itself was still generally carried out by the typists within the PQS organisation or outsourced to a specialist agency.

As personal computers began to become established from the mid-1980s, the tasks of calculating the quantities and placing into bill order could be done electronically by using coding systems. Measurement itself started to be carried out on-screen instead of longhand on paper, and the growth and development of computer-aided taking-off packages have continued to the present day.

Figure 5.1 shows a sample page from a bill of quantities, set in elemental format as the blockwork, plasterwork and paintwork are shown together since they all relate to the internal walls. If the BOQ was structured in trade sections, the blockwork would be in quite a different section from the plasterwork or painting (section F and section M).

One of the objectives of the Uniclass system is to provide a standard classification system across the whole construction industry. Therefore designers using CAD (computer-aided design) systems would code their drawings and details using the same system, and some of these CAD systems can produce bills of quantities automatically. However, this also means that the self-checking system has been bypassed, and the quality of the output is only as good as the quality of the input.

5.4.2 Bills of approximate quantities and schedule of rates

Bills of approximate quantities and schedules of rates are structured and prepared in exactly the same way as a firm BOQ, except that with the bill of approximate quantities, the quantities stated are not guaranteed and will need to be subject to remeasurement when the work is carried out on site. The purpose of putting an approximate quantity in the tender documents is to give the contractor an indication of the extent of the item required, so that any economies of scale can be calculated. Clearly, the unit rate (in £/m^3) will be lower if the project requires 1,000m^3 than if it only requires 5m^3. It is remarked in another chapter that civil engineering projects prepared under the ICE conditions of contract are 'remeasurement contracts', meaning that the quantities in the BOQ are not guaranteed and must be measured separately when the work is carried out on site. Therefore not only do the BOQs give only an indication of the scope of the item, but in the concrete work section of CESMM3, the contractor is also given an indication of the total amount of concrete across all items. In civil engineering projects, concrete work represents a significant proportion of the costs, and giving the

Table 5.1 Uniclass building elements and work sections

Section G: building elements		Section J: work sections for buildings	
G1	Site preparation	JA	Preliminaries/general conditions
G11	Site clearance	JB	Complete buildings/structures/ units
G12	Ground contouring		
G13	Stabilisation	JC	Existing site/buildings/services
G2	Fabric: complete elements	JD	Groundwork
G21	Foundations	JE	In situ concrete/large precast concrete
G22	Floors		
G23	Stairs	JE0	Concrete construction generally
G24	Roofs	JE1	Mixing/casting/curing/spraying in situ concrete
G25	Walls		
G26	Frame/isolated structural members	JE2	Formwork
G3	Fabric: parts of elements	JE3	Reinforcement
G31	Carcass/structure/fabric	JE4	In situ concrete sundries
G32	Openings	JE5	Structural precast concrete
G33	Internal finishes	JE6	Composite construction
G34	Other parts of fabric elements	JF	Masonry
G4	Fittings/furniture/equipment (FFE)	JF1	Brick/block walling
G41	Circulation FFE	JF2	Stone walling
G42	Rest, work FFE	JF3	Masonry accessories
G43	Culinary FFE	JG	Structural/carcassing metal/timber
G44	Sanitary, hygiene FFE	JG1	Structural/carcassing metal
G45	Cleaning, maintenance FFE	JG10	Structural steel framing
G46	Storage, screening FFE	JG11	Structural aluminium framing
G47	Works of art, soft furnishings	JG12	Isolated structural metal members
G48	Special activity FFE	JG2	Structural/carcassing timber
G49	Other FFE	JG3	Metal/timber decking
G5	Services: complete elements	JH	Cladding/covering
G50	Water supply	JJ	Waterproofing
G51	Gas supply	JK	Linings/sheathing/dry partitioning
G52	Heating/ventilation/air conditioning (HVAC)	JK1	Rigid sheet sheathing/linings
G53	Electric power	JK2	Timber board/strip linings
G54	Lighting	JK3	Dry partitions
G55	Communications	JK4	False ceilings/floors
G56	Transport	JL	Windows/doors/stairs
G57	Protection	JM	Surface finishes
G58	Removal/disposal	JN	Furniture/equipment
G59	Other services elements	JP	Building fabric sundries
G6	Services: parts of elements	JQ	Paving/planting/fencing/site furniture
G61	Energy generation/storage/conversion		
G62	Non-energy treatment/storage	JR	Disposal systems
G63	Distribution	JS	Piped supply systems
G64	Terminals	JT	Mechanical heating/cooling/ refrigeration systems
G65	Package units		
G66	Monitoring and control	JU	Ventilation/air-conditioning systems
G69	Other parts of services elements		
G7	External/site works	JV	Electrical supply/power/lighting systems
G71	Surface treatment		
G72	Enclosure/division	JW	Communications/security/control systems
G73	Special purpose works		
G74	Fittings/furniture/equipment	JX	Transport systems
G75	Mains supply	JY	Services reference specification
G76	External distributed services	JZ	Building fabric reference specification
G77	Site/underground drainage		

Example of a page from a Bill of Quantities

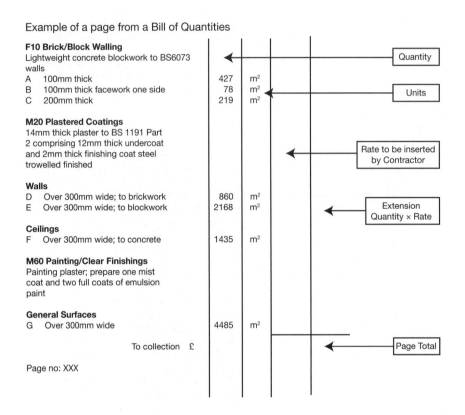

5.1 Sample page from a bill of quantities

contractor this kind of indication will help them make the decision whether to buy in ready-mixed concrete or install a batching plant to produce their own concrete locally.

A schedule of rates goes one step further in that there are no quantities included and the tendering contractor is expected to guarantee a rate that will be used whether 1,000m³ or only 5m³ are required. Clearly this does not allow for any economies of scale to be given and should be used only for relatively small projects or as part of a term contract where the contractor is required to carry out work at very short notice.

5.4.3 Activity schedules

An activity schedule is a list of the activities that the contractor expects to carry out in completing their obligations on the project. It is only relatively recently that this has been used as a pricing document and clearly there are advantages in using a time-/ programme-based payment mechanism rather than a quantity-based mechanism as with the BOQ. When the activity schedule has been priced by the contractor, the sum for each activity or each group of activities is the price to be paid by the employer for that activity or group. The total of all the activities and groups is the contractor's total price for carrying out the works (i.e. the contract sum).

A contract based on an activity schedule is essentially a lump-sum contract. When preparing their tender the contractor must consider the full scope of works and break this down into a number of identifiable activities in order to price each activity. This is usually known as a 'work breakdown structure' (WBS). In many cases, the tender documents will contain an outline activity schedule from the client in the employer's requirements, and the contractor is expected to develop this together with the WBS into a more detailed and priced activity schedule

It is essential that the activity descriptions are clear, unambiguous and complete so that the entire works are included within the overall activity descriptions and the work included within any one particular description can be readily identified. Since payment is usually made by the client only on completion of each activity or group and not before, each activity description should include a definition of the measure to be adopted to confirm completion – e.g. signing off by the project manager or contract administrator.

This form of payment mechanism is adopted in many standard forms of contract, as it can significantly reduce the administration burden. It is used particularly in design and build contracts where the contractor has control over the definition of the project, but can also be used effectively in more traditional procurement routes.

The NEC3 contract uses an activity schedule as the payment mechanism in options A and C, and internationally, the FIDIC Yellow Book (for plant and design build projects) and Silver Book (for EPC/Turnkey projects) at clause 14.4 allows payments to be made by instalments against a schedule of payments, which may be defined by reference to actual progress. This would therefore allow an activity schedule to be adopted.

Figure 5.2 gives an example of an activity schedule allowing the contractor opportunity for pricing the scheduled activities.

The advantage of using a schedule such as this as a pricing document is that the contractor does not have to convert their primary resource costs (labour, plant and materials) into the unit rates of a BOQ, which makes the costing of variations and delays much more accurate. The main disadvantage is that if the contractor shows their actual primary costs in this spreadsheet, together with the total cost of the activity, they have effectively informed the client of their overhead and profit mark-up, which is normally commercially sensitive and confidential information that they may well be unwilling to release. However, as mentioned in Chapters 9 and 10, in many partnering and framework arrangements, the contractors' costs are considered to be 'open book', where the percentage mark-up for profit and overheads is known to all parties to the project.

5.5 Summary and tutorial questions

5.5.1 Summary

The procedures for tendering and payment are very closely linked to procurement routes, as the choice of procurement will depend on the completion of the design at the point of contractor selection. An incomplete design should not use a lump-sum payment mechanism, since the scope of works is not yet established, so consequently the price cannot be established. The three methods of payment included in this chapter are:

Duration	Preceded by	Labour Cost	Material Cost	Plant Cost	TOTAL Cost	Aug 3	10	17	24	31	Sept 7	14	21	28	Oct 4	12	19	26	Nov 2	9	16	23	30	Dec 7	14	21	28
6 days																											
6 days	1																										
20 days	2																										
20 days	2																										
30 days	4																										
6 days	5																										
21 days	6																										
6 days	7																										
20 days	7																										
25 days	7																										
30 days	10																										
15 days	10																										
15 days	10																										
10 days	10																										
15 days	10																										
21 days	9																										
3 days	all																										

5.2 An example of a priced activity schedule

1   Lump-sum or fixed-price contracts, where the total cost of the works is fixed as a lump sum or single figure, which may be amended by client-generated changes. The lump sum is calculated as an aggregate of the cost of the items in the pricing document (normally a BOQ).
2   Firm-price contracts, where the items in the BOQ or schedule of rates will have unit rates inserted by the contractor and the quantities will be remeasured during the construction stage to assess the total cost of the works.
3   Cost-reimbursable contracts, where the contractor will be paid their costs plus an allowance for their own overheads and profit.

Each of these methods has its own advantages and disadvantages depending on the client's objectives in terms of the cost, time and quality criteria of the project.

As construction projects can range from small (generally considered as under £250,000) to very large (over £10 million) and the execution of the projects is very capital intensive (in terms of the requirement of major plant and equipment), the tendering process by construction firms is very expensive. Therefore, it is essential that this tendering process is conducted fairly and in accordance with good practice. In the UK and many other mature construction markets, this has developed into a sophisticated process covering prequalification, tender compliance, post-tender negotiation, award and mobilisation.

Finally, the pricing documents used for payment purposes can be classified as:

a   firm bill of quantities
b   bill of approximate quantities
c   schedule of rates
d   activity schedules.

Each of these pricing documents will again have its own advantages and disadvantages depending on the procurement route chosen for the project.

5.5.2 Tutorial questions

1   What do you understand by a 'fixed-price contract' and a 'firm-price contract'?
2   What would be the significant differences between tender documents and contract documents?
3   What are the advantages of having an order of priority within the list of contract documents?
4   Do you consider that open tendering would result in more competitive bids than selective tendering? Answer this question from the point of view of a client and a contractor.
5   What is a qualified tender?
6   Why are letters of intent dangerous?
7   What are the advantages of elemental bills of quantities compared to bills of quantities produced in work sections?
8   What are the advantages of activity schedules as a pricing document?
9   Should activity schedules be produced by the client's consultants or the contractor?

6   Separated procurement systems

The main characteristic of this approach is the separation of the responsibility for the design of the project from that for its construction. It contains one major procurement system – the *traditional system*, and, as mentioned in Chapter 5, can be either tendered in open competition, or from a selected list of pre-qualified contractors. The system can be further sub-divided to enable the successful contractor to be involved in the design stage (two-stage tendering) or by a process of negotiation, the contract sum is established by agreement rather than by competitive tendering.

6.1 'Traditional' system: single-stage lump-sum competitive tendering

This method of procuring building projects is usually referred to within the industry and literature as 'the traditional method', although it has only been traditional for a relatively short time. Up to about the middle of the nineteenth century, buildings were generally procured by a master craftsman employing other tradesmen who were paid direct by the client (we would call this construction management in the modern terminology). It is therefore only for about 150 years that buildings have been built using a main contractor, with the design and supervision being carried out by an architect assisted by other specialist consultants. The term 'traditional method' has, therefore, been used to describe this system throughout this book.

As well as the separation of design and construction, the traditional procurement system has other characteristics:

1   Project delivery is a sequential process. See Figure 6.1.
2   The design of the project should be completed before work commences on site.
3   The responsibility for managing the project is divided between the client's consultants and the contractor, and there is therefore little scope for involvement of either of the parties in the other's activities. See Table 4.2 in Chapter 4, section 4.10.1.
4   Reimbursement of the client's consultants may be on a fee and expenses basis, although this is becoming increasingly rare, whereas the contractor is paid for the work completed on either a remeasurement or lump-sum basis.

The purest form of the traditional procurement system will include all of these features, although nowadays any of the pure forms of construction procurement are few and far between.

At the beginning of the project, the client appoints independent professional consultants, who fully design the project and prepare tender documents upon which

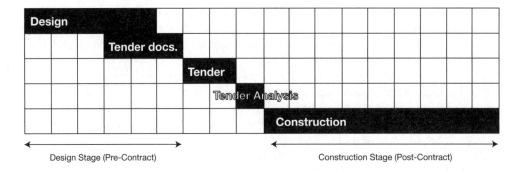

6.1 The traditional procurement system in simplified diagrammatic form

competitive bids, often on a lump-sum basis, are obtained from interested construction firms. The successful tenderer enters into a direct contract with the client and carries out the work under the supervision of the original design consultants.

This traditional procurement system has been used by the majority of clients of the industry for at least the past 150 years, as the system came to prominence in the early nineteenth century, when, as a result of the frequent disputes between clients and the separate tradesmen involved in the earlier 'measure and value' system, it was felt that a change was needed, and as ever, government took the initiative when the Office of Works introduced competitive tendering for entire projects as a superior alternative.

By the 1850s, 'contracting in gross', as the system was then known, widely prevailed, enabling clients to secure the economic benefits of competition, knowledge of the final cost before work began, better control of subsequent expenditure and the ability to enter into a single, contractual relationship with a builder instead of multiple tradesmen, who were able to co-ordinate work between themselves. At the same time as the organisations for implementing projects and associated contractual relationships were changing, surveyors were being used by groups of builders, and eventually by clients, to take off the quantities of materials required for estimating purposes from the architect's drawings, thus relieving the tenderers of responsibility for their accuracy and sufficiency. And so began the development of the quantity surveying profession.

From the second half of the nineteenth century until the Second World War, the use of the traditional system by both public and private clients gradually increased to the point where, in England and Wales at least, this method was being used for the vast majority of projects, albeit on the basis of very open competitive tendering.

After the end of the war in 1945, the traditional procurement system remained predominant, supported by the findings of the Simon Committee, which had reported that an examination of the system of placing contracts had not disclosed any serious weakness in the methods that had been built up by 'architects, quantity surveyors and builders', although it was recommended that, where appropriate, selected and limited lists of tenderers should be used together with negotiation (Ministry of Works 1944).

This method of building procurement continued until the early 1960s, with the vast majority of post-war reconstruction in the UK organised in this way. Because there was so much work to do and nobody could think of a better way, very little thought was given to alternative methods of delivery. The major disadvantage of the traditional system is the fragmentation of responsibility. Clearly it was better than having lots of different tradesmen milling around, but the lack of co-ordination between design and production was starting to cause major problems. Additionally, as the plumbing and electrical services in buildings became more complex, the need for co-ordination across the trades was increasingly important.

The seeds of change were sown in the Emmerson and Banwell Reports of 1962 and 1964. Emmerson concluded that ways needed to be found to improve co-ordination and co-operation among the building owner, consultants, contractors and subcontractors, and suggested that the system for placing contracts and managing projects should be comprehensively reviewed.

Banwell reiterated this view and berated the majority of the various members of the industry for their reactionary approach to new ideas and processes, pointing out the urgent need for the separate factions to come together and think and act as a whole, particularly in the 'letting, form and management of contracts'.

Perhaps as a result of the pressure exerted by such reports, as well as by normal commercial influences, the mid- to late 1960s were a watershed in terms of the increased use of non-traditional procurement systems and thus the decline in the employment of traditional methods.

This period also saw the beginning of the growth of design and build, the birth of the first major management contract, the use of serial tendering for system build/ industrialised buildings and a general acceptance among the larger and more far-sighted clients and consultants that the involvement of the contractor at an early stage was a good thing and could be of great benefit to the project as a whole.

In 1973, the international oil crisis, and the consequent major rise in fuel prices and interest rates, meant that the dominant project objective of most clients became the need for rapid commencement and completion of a development in order to reduce borrowing to a minimum. These requirements resulted again in a continuing decline in the use of traditional procurement systems in favour of systems which would allow the building to start as quickly as possible, even though the design may not yet be fully developed.

This trend has continued until the present time, with clients satisfying their needs by increasingly using non-traditional procurement systems to the detriment of the traditional method.

While a great deal of time and effort has been expended in achieving technical innovation in construction and refining existing and producing new forms of contract, very little attention had been paid, until comparatively recently, to the rationalisation and reorganisation of the procurement process, thus allowing the traditional system to maintain a major share of the available building work.

However, the increasing participation of large and sophisticated clients in this process, particularly during the past twenty years or so and the development of the Private Finance Initiative in the mid-1990s (see Chapters 3 and 11) has meant that the amount of work being carried out using the traditional procurement method has declined, a trend that appears likely to continue.

There is a wide variation in the estimates of the level of use of this system but looking at Tables 3.1 to 3.6 in Chapter 3 shows that the lump-sum contracts with or without BOQs represented almost 90 per cent of activity by number of contracts throughout the late 1980s, which reduced to 74 per cent by 2004. This reduction was caused primarily by an increase in design and build procurement from 3.6 per cent in 1985 to 13 per cent in 2004, peaking at over 20 per cent in 1998.

A slightly different picture emerges if we look at the procurement methods in terms of value of contracts. Traditional procurement represented 70 per cent of construction in 1985 and reduced dramatically in 2004 to 34 per cent, which corresponded with a massive increase in design and build. Therefore, more of the higher-value contracts are being let as design and build (presumably by experienced clients), leaving the smaller and medium-sized contracts as traditionally procured (by less experienced clients). The experienced clients are therefore more willing and able to manage non-traditional procurement.

At the present time, the market's perception of the traditional system and its variants is that for small and medium-sized uncomplicated projects where time is not at a premium and where the client wishes to have continuous control over the design, this system offers the most economical route to success. In the case of major new works, where time is at a premium or the number of competent contractors is low, the shortcomings of the separated methods are likely to result in the use of non-traditional systems.

6.1.1 The process

The outline plan of work drawn up by the Royal Institute of British Architects (RIBA) sets out the process to be followed for projects being procured by traditional means.

Although the plan identifies twelve stages, there are only four which need concern us in the context of procurement systems. These are *preparation*, *design*, *tender* and *construction*.

As one of the unique characteristics of the traditional system is that it follows a strictly sequential path, each of these four stages can be viewed as separate entities and carried out, to a certain extent, in isolation from the others, with the result that the process can become extremely lengthy, lead to poor communication, undermine relationships between project-team members and produce problems of buildability. The four stages are now described and discussed.

6.1.2   Preparation for the project

This is the inception stage of the project, when the client establishes their needs in principle within their outline business plan, appoints a project manager or other lead consultant and selects and appoints a design team, which normally consists of, as a minimum, an architect, a structural/civil engineer, a mechanical and an electrical engineer and a quantity surveyor, together with any other specialist consultants necessary for the successful implementation of the project. Depending upon the nature of the client, and the undertaking itself, the project manager may be:

- an employee of the client organisation with little or no construction knowledge who simply acts as a co-ordinator of information and a single source of contact and communication for the design-team leader, who will have responsibility for the day-to-day project management;
- an experienced construction professional, permanently and directly employed by the client, who, in addition to being the single point of contact, will be responsible for the financial, technical and administrative management of the project from inception to completion;
- an external consultant project manager appointed on a professional services contract for a specific project to carry out the same duties as an 'in-house' project manager.

Stage	Purpose of work and decisions to be reached	Tasks to be done	People directly involved	Commonly used terminology
A Inception	To prepare general outline of requirements and plan future action.	Set up client organisation for briefing. Consider requirements, appoint architect.	All client interests, architect.	Briefing
B Feasibility	To provide the client with an appraisal and recommendation in order that he may determine the form in which the project is to proceed, ensuring that it is feasible, functionally, technically and financially.	Carry out studies of user requirements, site conditions, planning, design and cost etc as necessary to reach decisions.	Clients' representatives, architects, engineers and QS according to nature of project.	
Stage C begins when the architect's brief has been determined in sufficient detail.				
C Outline Proposals	To determine general approach to layout, design and construction in order to obtain authoritative approval of the client on the outline proposals and accompanying report.	Develop the brief further. Carry out studies on user requirements, technical problems, planning, design and costs, as necessary to reach decisions.	All client interests, architects, engineers, QS and specialists as required.	Sketch plans
D Scheme Design	To complete the brief and decide on particular proposals, including planning arrangement appearance, constructional method, outline specification, and cost, and to obtain all approvals.	Final development of the brief, full design of the project by architect, preliminary design by engineers, preparation of cost plan and full explanatory report. Submission of proposals for all approvals.	All client interests, architects, engineers, QS and specialists and all statutory and other approving authorities.	
Brief should not be modified after this point.				
E Detail Design	To obtain final decision on every matter related to design, specification, construction and cost.	Full design of every part and component of the building by collaboration of all concerned. Complete cost checking of designs.	Architects, QS, engineers and specialists, contractor (if appointed).	Working drawings
Any further change in location, size, shape, or cost after this time will result in abortive work.				
F Production Information	To prepare production information and make final detailed decisions to carry out work.	Preparation of final production information ie drawings, schedules and specifications.	Architects, engineers and specialists, contractor (if appointed).	Site operations
G Bills of Quantities	To prepare and complete all information and arrangements for obtaining tender.	Preparation of Bills of Quantities and tender documents.	Architects, QS, contractor (if appointed).	
H Tender Action	Action as recommended in relevant NJCC Code of Procedure for Selective Tendering.	Action as recommended in relevant NJCC Code of Procedure for Selective Tendering.	Architects, QS, engineers, contractor, client.	
J Project Planning	To enable the contractor to programme the work in accordance with contract conditions; brief site inspectorate; and make arrangements to commence work on site.	Action in accordance with RIBA Plan of Work.	Contractor, subcontractors.	
K Operations on Site	To follow plans through to practical completion of the building.	Action in accordance with RIBA Plan of Work.	Architects, engineers, contractor, subcontractors, QS, client.	
L Completion	To hand over the building to the client for occupation, remedy any defects, settle the final account, and complete all work in accordance with the contract.	Action in accordance with RIBA Plan of Work.	Architects, engineers, contractor, QS, client.	
M Feedback	To analyse the management, construction and performance of the project.	Analysis of job records. Inspections of completed building. Studies of building in use.	Architects, engineers, QS, contractor, client.	Feedback

6.2 RIBA Plan of Work

Therefore the traditional procurement system involves the client in a number of differing relationships with several organisations, and many inexperienced clients are often dismayed at the complexity of the process and the size and cost of employing the design team itself before any physical progress is apparent.

During the preparation stage and before the appointment of the design team, but not the project manager, the client establishes their basic needs in terms of the purpose and quality of the project together with the cost and time parameters. Having settled these fundamental requirements, the strategies that will be used to implement the project successfully will be determined and, on the basis of these, an appropriate design team will be appointed. This is the basis of the briefing stage, where the employer's requirements are established from the client's outline business plan and a formal brief is written so that a design solution can be developed.

It is important during this stage to take time to ensure that the client's requirements are correctly established, and this is often time well spent, which is subsequently reflected in the ability to proceed with the other phases of the project with the minimum of change and disruption. The decisions taken at this time set the whole tone and pattern for the remainder of the building process. Unfortunately, in most cases, commercial necessities mean that the project is required on site as soon as possible, which is often the source of conflict when using the traditional system.

At this stage, and to a slightly lesser extent during the design phase of the project, the client has a great deal of influence – they have much less opportunity to control any aspects of the undertaking during the last two stages and particularly during the construction period.

6.1.3   The design stage

Before any design work can be started, the client is under a legal obligation to appoint a CDM co-ordinator to oversee the health and safety issues during the design stage. The design team can now be appointed, which will develop the project through a series of sub-stages:

- detailed briefing
- feasibility
- outline design
- scheme design
- detailed design.

The scheme's configuration and features become firmer at each stage. Again, the client and their consultants have considerable freedom during this phase to conceive and develop the project without excessive time or economic pressures, although projects which move quickly in the preconstruction period tend to be constructed quickly as well.

During the design process, the designers are often working in isolation, far removed from the contractor who will eventually be responsible for carrying out the construction of the project and sometimes from each other. This isolation from the contractor is deliberate and great care is taken to ensure that no contact occurs. In fact, the actual constructor has not yet been appointed and will only be appointed by competition from those who submit tenders. However, it does serve to reinforce the division between the role of designer and constructor.

As a consequence, opportunities for ensuring that the design solution can actually be built economically and efficiently (this concept is known as 'buildability') are virtually non-existent, although in some cases the appointed builder may suggest design modifications during the tender or negotiation stages before work starts on site. In this case, it is vital that the new modified design/scope of works is incorporated into the contract documents.

The main reasons for the lack of involvement of contractors when using the traditional method are well known but nevertheless deserve reiteration:

1   Clients wish to ensure that the responsibility for the design of any project is vested in one group, i.e. the design consultants.
2   The list of tenderers will generally not be available until the design has been largely completed.
3   The practical and ethical difficulties of dealing with suggestions from a number of contractors during the design stage are difficult to overcome.
4   Once design decisions have been made, depending on their complexity and strategic value, they usually cannot be changed without cost implications and possible delays to the construction programme. For example, changing the foundations from raft to piles would involve a major redesign of many other elements.

Although the client may be anxious to see work commence on site, progress during this design stage should be carefully controlled and not unreasonably forced. Hastily prepared design details can lead to mistakes, ambiguities and misunderstandings and hence disputes during the construction stage, which generally result in delays and additional unnecessary costs to the client.

6.1.4 Preparing and obtaining tenders

Tender documentation on traditionally procured projects normally consists of:

* tender drawings
* specification(s)
* a pricing document (usually a bill of quantities, but may be an activity schedule).

The bill of quantities is prepared by the consultant quantity surveyor on the basis of measurements 'taken off' the designers' drawings in order to provide each tenderer with a common base from which to price their bid. For the traditional system to operate successfully, and to minimise the financial risk to the client, it is imperative that the design is fully developed before the bills of quantities are prepared and tenders invited. If this is not done, excessive variations and disruptions of works are likely to occur.

Although selection of the contractor by limited competitive tendering should offer the assurance of achieving the lowest price for the project, in reality the designers' drawings are rarely in sufficient detail to enable a bill of quantities to be prepared with 100 per cent accuracy, and the art of evaluating from the drawings the exact amount of work required presents challenges which range from the difficult to the impossible, but most PQS firms would admit to being experienced in this issue.

The selection of contractors who will receive tender documents and submit bids can be made in a number of ways, but it is usual in the case of traditionally managed projects for tenderers to be selected from a list of approved firms, who have pre-qualified to be allowed onto the tender list. It is generally only the occasional project and those governed by European Union regulations where an open advertisement invites contractors to bid.

The main difficulty with open tendering is that the client or their consultants would have to send a complete set of tender documents to every potential bidder who requests them, whether the bidder is serious about submitting a tender or not. Clearly, this can be a very expensive exercise, and where open tendering is used, it is not uncommon for a fee to be charged for the documents, returnable on submission of a bona fide bid. This is designed to deter spurious requests for the documentation from parties who may be curious about the project but have no intention of submitting a bid.

Therefore, as described in Chapter 4, the selection of the contractor for the works is generally made by selective tendering based upon a list of tried and tested contractors whose past performance, financial stability and resources have already been established and, in most cases, regularly monitored.

A central element of the tender documentation is the form of contract, which includes the terms and conditions governing the relationship between the client and the contractor. There are many standard forms of contract in existence, and it is normal and advisable to use the most appropriate standard form of contract for that particular project and procurement route. The benefits of using a standard form are that both parties know what is included and what is not included and most importantly the rights and responsibilities of both parties. In many projects, one of the parties (usually the client) wishes to vary or amend the clauses in the standard form in order to transfer more of the risk to the other party. This is not usually a good idea, since most of the clauses are mutually dependent, and amending one clause may have unintended consequences somewhere else.

If for any particular reason it is intended to introduce any special conditions of contract, to use non-standard agreements proposed by one of the parties or, more rarely, to formulate a bespoke form of contract, technical and legal advice is essential and the advantages and disadvantages of the proposed documentation should be carefully considered before any commitment is made.

Once tenders are received, the selection of the best bid is fairly straightforward if it is judged on price alone, having been based on documentation which is common to all tenderers and which, theoretically at least, accurately and comprehensively reflects the client's actual requirements. However, care must be taken at this point, as the bids will often be only a single lump sum. For example, the form of tender may state only:

We, XYZ Construction Company Ltd. Offer to carry out the Works known as xxxxx in accordance with the tender documents, for the sum of £876,543.00.

This may be the lowest bid, but how do we know it is a serious bid that has been properly estimated? It is therefore important to conduct a tender analysis, where the lowest two or three tenders are called in for technical and arithmetical checking. The National Joint Consultative Committee (NJCC) for the construction industry, which was made up of representatives from all sides of the industry, published a series of documents on the Code of Procedure for Tendering and although the NJCC no longer exists as a consultative committee, the codes are still an extremely useful guide to how tenders should be managed, evaluated and chosen. The Code of Procedure for Selective Tendering recommends that the three lowest tenders are checked before finally selecting the successful contractor.

When using the traditional system of procurement, the successful contractor will need an adequate period of time to plan the project thoroughly and organise the required resources. Undue haste in making a physical start on site may result in managerial and technical errors being made by both the design team and the contractor, which could lead to a lengthening rather than a reduction of the construction period.

A major disadvantage of the traditional route is that a very high proportion of the estimated cost of the project has been committed before work commences on site, although actual expenditure is comparatively small. This is because costs are generated by design decisions, but they are incurred when the work is carried out. However, it is during the construction phase that the majority of difficulties will surface, with the quality of the performance during this period having already been largely determined by the quality of design decisions. It is at this stage that the price for an incomplete design, inaccurate bill of quantities, poorly prepared tender documentation and lack of 'buildability', etc. is paid. For example, the design may stipulate that pad foundations are to be used below the columns. Clearly, these pad foundations incur a cost, in excavation, disposal, earthwork support, concrete, reinforcement, formwork, etc. Therefore these costs have been generated by the decision to use pad foundations, but the costs will only be incurred when the work is carried out. However, because the design stage was rushed, the detailed ground conditions were not established prior to the decision to use pad foundations, and the building now requires piled foundations with pile caps below the superstructure columns. The actual costs will be significantly more than originally planned and the construction period will be lengthened while the piled foundations are designed and constructed.

The ability to introduce changes to the design of the project during the construction period is a characteristic of both the British construction industry and the traditional procurement system in particular. Of course, this ability to add work to and omit work from the project has both advantages and disadvantages in that the client or architect can add work in during the construction stage that they 'forgot' at design stage, or omit work in order to reduce the cost of the project. Also, such variations have been identified as one of the most important causes of delay, so if it is essential to instigate changes, the project team (including the contractor) needs to be consulted and both the practical and financial consequences of the proposed variation established in detail before instructions to proceed are given. The nature of lump-sum contracting is that any variation to the contract documents will require a formal variation order from the contract administrator, which includes any issues for which they are responsible, which often causes political sensitivities during the construction stage. Issues such as these, which are outside the contractor's control, may be subject to extensions of time and loss and expense payment to the contractor.

The management and supervision of the work on site to ensure that it conforms to the client's brief as reflected in the design, specification and contract conditions are generally the responsibility of the design team, although it should be remembered that under normal terms of engagement the design team is not required to carry out full-time supervision of the works. This is usually an additional service provided by a resident architect, engineer and/or clerk of works, who is employed at the client's expense.

The ability of design consultants in general, and architects in particular, to manage projects has been continually questioned over the past three decades, when it has been maintained that in the case of the traditional approach designers are not motivated to give sufficient attention to the control of the critical criteria of cost and time and have not been trained to manage such projects effectively.

The combination of part-time supervision and lack of management expertise and motivation during the construction phase of traditionally procured projects often results in delays and additional costs being incurred by the client as a consequence of poor performance by their consultants. The detailed and continuing involvement of the client can offset these deficiencies, as it has been amply demonstrated that customers who take a constructive and objective interest in all aspects of their projects achieve the best results, particularly in terms of speed of completion.

Because of the separated nature of the traditional method of procurement, it is necessary for the client to ensure that good communications exist between all members of the project team, so that immediate decisions are made when queries arise during the construction phase and so that a strong site-management team is in place before work commences on the project.

Payment to the contractor for work that has been satisfactorily completed is made by means of interim certificates, generally monthly, but may also be on the basis of activities completed, to the value of work done, issued by the architect or contract administrator on the recommendation of the quantity surveyor. The priced bill of quantities submitted by the contractor at tender stage forms the basis of these interim valuations and also ensures that any variations can be valued by reference to pre-agreed rates for appropriate items of work. An agreed percentage is usually retained until practical or substantial completion of the works, when a portion (usually half) is paid to the contractor. The remaining half of the retention fund is paid at final completion after the defects correction period. However, it is now becoming more common for contractors to purchase a 'retention bond' from a bank, which does away with the need for the clients to retain a proportion of the interim valuations and thereby improves the contractor's cash flow – it is their money anyway.

The Latham Report (HMSO 1994a) considered the issue of payments in some detail, and following this report, the Housing Grants, Construction and Regeneration Act 1996 stipulated that all construction contracts must make provision for interim payments and that any payment duly certified must be paid to the contractor within a set time unless a formal withholding notice is given. It also stated that 'pay-when-paid' clauses are illegal, so that a main contractor cannot refuse to pay a subcontractor merely because they have not received the payment from the client. All this signifies a resolution by the government and the courts to improve the cash flow in the construction industry and that more of the contractor's efforts are effectively engaged in actually managing the project, rather than pursuing outstanding payments, and should serve to ensure good relations and thus improve project performance.

The sequential characteristic of the traditional system can also mean a reduced ability to deal with any unexpected delays. Overcoming such delays during the construction phase of the project is not easily achieved, even if the cost of the necessary acceleration can be accommodated within the financial budget for the scheme, and the project team needs to monitor closely the contractor's progress so that any areas of possible delay can be detected sufficiently early to enable remedial action to be taken and practical completion achieved in accordance with the client's requirements.

The major selling point of the traditional system and one which is made by countless consultants across the world is that, by using bills of quantities as part of the tender documentation, the cost of tendering is reduced, the quantitative risks encountered in tendering are removed, competition is ensured, post-contract changes can be implemented at a fair and reasonable cost and clients can be confident that they know their financial commitment. All this is true, provided that the design has been fully developed and accurately billed before obtaining tenders. If, however, these criteria have not been strictly met, excessive variations, disruption of the works and a consequent increase in the tendered cost will occur. There is often in addition a considerable amount of work covered by provisional sums which will necessarily vary when the actual extent and cost of the work is known, so the final account may be quite different from the original tender lump sum or contract lump sum.

Because of its sequential nature, the traditional procurement system is generally regarded as the slowest method of procuring construction projects available to a client, which may be mitigated by the use of variants of the system during the preconstruction phase. Further savings could be made during the construction phase itself by overlapping some design work into the construction phase by the use of subcontractors with a design element, but this would accentuate the problem of co-ordination, mentioned earlier.

6.1.5 *Quality and functional suitability*

The generally held view among clients is that the traditional procurement system provides a high degree of certainty that quality and functional standards will be met. This view has been supported by the findings of various research projects which examined the existence of faults in completed buildings when designed by different designers working within various procurement systems. For example, on factory buildings, designed by architects who were employed directly by clients, almost half were found free of faults, 40 per cent had minor defects and 4 per cent had major problems. In the case of office buildings designed under the same circumstances, 54 per cent had no faults, 41 per cent had minor defects and 5 per cent had major defects.

Most research in this area identifies the traditional procurement method as being suitable for projects with normal to more demanding levels of building quality. As and when the construction industry comes out of the latest recession, clients will naturally revert to this method of procurement, as it is seen as giving value for money on account of the competitive nature of the tendering process.

6.1.6 *Other characteristics*

The traditional procurement system has the advantage of having stood the test of time over many years and being understood by many clients and by all the participants from the construction industry itself. The client is able to select the most appropriate design team for their project, taking advantage of designers' experience of similar developments, and can also delay commitment to a building contract until a later stage in the development of their requirements, which is an extremely useful benefit given the uncertainties in the financial markets and banking sectors.

The main disadvantages of the system have already been highlighted, but there is little doubt that the traditional procurement method does not motivate the client to make decisions as firmly, or as early, as they possibly should do, nor does it encourage designers and contractors to pay enough attention to saving and controlling cost, or time, and improving building quality. It is also the case that the designers of the project often have no direct experience of managing construction work and the contractor is unable to contribute to the design of the project until it is perhaps too late. Additionally, most standard forms of contract for this route are predominantly adversarial in nature and therefore do not encourage collaborative or co-operative working, but rather a silo mentality of 'that's not my job'.

However, it should be said at this point that enormous pressure is usually needed to change procedures, such as the traditional procurement method, which become engrained and institutionalised within the industry, but since the mid-1990s such pressure has indeed been exerted, mainly by the government and large property developers, to the point where some would argue that non-traditional procedures have been used inappropriately and the traditional system discarded unnecessarily – possibly a case of throwing the baby out with the bathwater.

What clearly needs to be sought is a balance between the use of proven non-traditional methods and the known ability of the traditional system to offer the most economic procurement route for small and medium-sized uncomplicated projects where time is not at a premium.

6.2 Two-stage selective tendering

This is a suitable method where the early involvement of the main contractor is required before the project is fully designed, as it enables the design team to make use of the contractor's expertise in construction planning and buildability. The contractor thus becomes involved in the planning of the project at an early stage, giving them more opportunity to influence the design by using their skills and knowledge of production methods. See Figure 6.3 for a diagrammatic illustration.

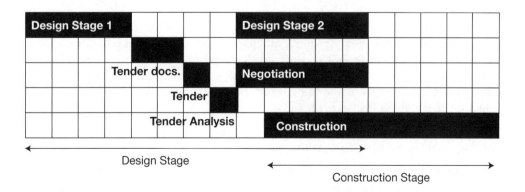

6.3 The two-stage procurement system in simplified diagrammatic form

The first stage of the two-stage system consists of the selection of the contractor on the basis of a competitively priced tender but with minimal information provided, because the project is still at outline proposal or scheme design stage (C and D in the RIBA *Plan of Work* in Figure 6.2). The tender submission is on a basis of the layout and design of the works, clear pricing documents relating to the preliminary design and specification and the conditions of contract.

In the second stage, the employer's professional team collaborates with the appointed contractor in the development of the detailed design for the whole project. A bill of quantities (or other pricing document) is prepared and priced on the basis of the first-stage tender. If an acceptable sum is produced the formal contract documents are then prepared and executed. This method is considered to be useful for building works of a large or complex nature, where the overall brief or client's requirements are unlikely to change. It is recommended for projects where the design and construction phases need to overlap because of time pressures and the contractor's design and production expertise can be utilised. For example, if the contractor had a particular expertise in bored friction piles and the ground conditions were particularly suitable for this type of foundation, it is clearly sensible to use the contractor's skills in designing the piling works. Moreover, if the project requires very careful phasing because the building is required to remain in use during construction works, the contractor's expertise in project planning would clearly be invaluable during the preconstruction phase.

6.3 Serial tendering

Serial tendering is used to obtain prices for a number of similar projects with the same client. The tendering contractors will bid at the beginning of the series of projects on the basis of a schedule of rates (i.e. with no quantities attached). The prices in this schedule of rates are then used in each of the series of projects, so that each project will be able to commence in a much shorter time than normal.

As far as the contractor is concerned, the serial tender is a standing offer to construct each project in the series for the rates in the original schedule. The series will usually consist of a minimum of three to perhaps twenty projects, depending on their size. The number will often be known at the time of tender, although more projects may be added by negotiation between the client and contractor. The precise quantities will be calculated as each project is designed and a separate contract may well be formed for each project in the series.

Serial contracts are ideal for programmes where the client knows that there will be a series of projects to be built of similar quality and specification but possibly different in size and location; however, the contractor's management and preliminaries should be roughly equal for all projects. It is also useful in obtaining firm prices from a contractor over a lengthy period of time and from the contractor's point of view it will hopefully mean guaranteed work.

6.4 Negotiated tendering/competitive dialogue

Clients may sometimes decide to negotiate a contract with one particular contractor rather than ask them to submit tenders based on designs, specifications, etc. There are several reasons for this: first, the client may have worked with the contractor before and been pleased with their performance; second, the design of the project may not be

Table 6.1 Key similarities and differences between competitive dialogue and negotiation

Competitive dialogue	Negotiation
Pre-qualification is on the basis of PQQ, *OJEU* notices and descriptive document of client requirements.	Pre-qualification is on the basis of PQQ, *OJEU* notices and descriptive document of client requirements.
Pre-qualified bidders participate in a staged dialogue involving proposal development based on the client's requirements, interim submission and iterative review.	Pre-qualified bidders respond to a fixed set of formal contractual deliverables and are required to submit a fully compliant bid.
Two-way confidential dialogue between the client's team and various bidders.	Indirect clarification of bidder queries is permitted with all responses copied to all bidders.
Bids based on different contractor's proposals, all of which are evaluated to confirm compliance with client requirements. Invitations to tender tailored to bidder's proposals. Competitive pressure maintained until winning tender selected.	All bids are based on the technical, legal and financial proposal described in the invitation to tender. Non-compliant bids are rejected.
Proposals progressively evaluated during a further dialogue stage. Some bidders may be eliminated using predetermined procedures.	–
Dialogue and opportunity to alter the proposals brought to a close before invitation of commercial tenders.	–
Bids are assessed on the basis of most economically advantageous tender only (not necessarily lowest).	Bids are assessed to identify the preferred bidder to take forward to the negotiation stage. The basis of assessment is either lowest price or most economically advantageous tender.
Post-bid discussions with tenderers and pre-award discussion with preferred bidders. Strictly limited to 'fine tuning'. No major changes permitted.	Extensive negotiations are permitted with preferred bidder. Potential for introducing change.
Contractual agreement established based on the preferred bidder's tender submission.	Contractual agreement established by best and final offer at end of negotiation stage.

sufficiently progressed to allow the contractor to submit a reasonable bid. Therefore, there are clearly significant risks to the client in negotiating with one contractor, but it all comes down to trust in the end.

Competitive dialogue was established in article 29 of the EU Public Sector Procurement Directive 2004 and in the UK by the Public Contracts Regulations (2006). It was designed to provide an alternative to the growing use of negotiation on complex projects, and to make better use of the private sector's role in delivering innovation.

The use of competitive dialogue is expected on projects where the client is able to state their requirement at the outset, but has not undertaken any design work. This need to keep options open may result from technical, legal or financial issues such as alternative design solutions, risk allocation arrangements, etc.

At first glance, competitive dialogue appears to require less preparation by the client because the development of the design solution is carried out by the bidders, but bodies such as the Office of Government Commerce (OGC) advise that thorough preparation should be undertaken, so that the client can fully brief participants and respond appropriately to the various bidders' proposals during the dialogue period.

Despite the substantial difference in process between negotiation and competitive dialogue, the outcome should be similar: an affordable and compliant preferred bid on which the parties can proceed to formation of contract.

The benefits from competitive dialogue are primarily related to the more detailed testing of the preferred proposal. In practice, the application of competitive dialogue has also revealed the following benefits:

- Both the client and the delivery partner (contractor) have greater confidence in the quality of the solution and the submission, particularly if it has been progressively tested during the dialogue process.
- Competitive dialogue does generate alternative design proposals, in the same way as design build, giving greater potential for added value in project delivery.
- The iterative process of design development in meetings between the bidder and the client means that the final building is more likely to achieve client satisfaction.

Dialogue is, however, an extended process. A typical three-stage dialogue, involving three sets of deliverables and assessments prior to the closure of the dialogue, could take around eighty weeks, excluding the client's initial development work. Total costs are also higher, with three bidders incurring significant bid costs, which, in most cases, the client is unlikely to pay for.

Clients also need to be aware that there is no binding offer on the table during the process until final bids are requested, which is similar to the financial close stage of PFI. Although the overall objective of the dialogue is to progressively develop proposals that are compliant and affordable, there is no pressure on a tenderer other than the competitive pressure from other bidders.

6.5 Summary and tutorial questions

6.5.1 Summary

The separated category has one fundamental and detrimental characteristic in that the responsibility for the two main elements of 'design' and 'construction' is vested in two quite distinct and separate entities – the design team and the contractor.

The advantages and disadvantages of the traditional procurement system are as follows.

Advantages

- Provided that the design has been fully developed and uncertainties eliminated before tenders are invited, tendering costs are minimised, proper competition is ensured, the final project cost will be lower than when using the majority of other procurement methods and the selection of the bid that is most advantageous to the client will present little difficulty.

- The existence of a priced bill of quantities enables interim valuations and progress payments to be assessed easily and variations or change orders to be quickly and accurately valued by means of pre-agreed rates.
- The use of this method provides a higher degree of certainty that quality and functional standards will be met than when using other systems.

Disadvantages

- Where tenders are obtained on the basis of an incomplete design, the bids obtained can be considered only as indicative of the final cost, and the client, who may well consider that the lump sum represents the actual final cost, is vulnerable to claims for additional financial reimbursement from the contractor.
- The sequential, fragmented and adversarial nature of this system can result in lengthy design and construction periods, poor communication between clients and the project team and problems of buildability.
- While the ability to respond to late demands for change, by introducing variations, can result in satisfied customers, such action has been identified as one of the main causes of delay and increased cost, and can lead to a permissive attitude to design changes.

6.5.2 Tutorial questions

1   At what points in the traditional procurement system would the following client decisions occur?

   a   the commitment to design the building
   b   the commitment to construct the building?

2   What kind of difficulties would be encountered in terms of co-ordination between the parties during the

   a   feasibility stage
   b   outline design stage
   c   detailed design stage?

3   Discuss the problems associated with the construction planning not being able to start until the tender stage.
4   What are the main advantages of two-stage tendering compared to single-stage tendering?
5   What are the main differences between serial tendering and framework agreements? You will need to refer to Chapter 10 for a discussion of framework agreements.

7 Project-specific procurement
Overlapping roles

Since the late 1980s, there has been a substantial increase in the use of management-orientated procurement systems in the UK, mainly as a result of clients, particularly those in the commercial sector, demanding earlier starts on site and a shorter overall project duration than could be achieved using conventional methods. They have also demanded more control over project costs and higher standards of functionality and quality than have been obtained previously by other means of building procurement.

There are two main forms of 'overlapping' procurement in use in the UK:

a   management contracting
b   construction management.

The term 'overlapping' refers to the roles carried out in construction projects (as defined in Chapter 4) being carried out concurrently, hence overlapping. In the separated procurement routes discussed in Chapter 6, the roles are quite discrete, both in terms of scope and when they are carried out in a project. In these systems, they overlap to a much greater extent, so that the design of later sections may not even have begun by the time that the groundworks and foundations are being executed.

7.1 Management contracting

Management contracting first started as a serious form of procurement in the early 1960s as a result of such changes as:

1   the increasing diversity, complexity and standardisation of building techniques;
2   the growing prominence of the specialist subcontractor;
3   the growth in size of projects, demand for tighter time and cost targets and for a more unified and purposeful management of the total process.

The systems that have been included within this category exhibit the characteristics common to all management-orientated systems where the contractor is included in the project team at the beginning of the design stage with the responsibility for the integration of the management of both design and construction. The purpose is to enable the project to better meet the client's needs, especially where accelerated commencement and completion are involved, than would be the case if the work was carried out using more conventional procurement systems.

7.1.1 Definition and history of management contracting

There are several definitions of management contracting, as the name is also used for particular systems of procurement offered by individual construction organisations, especially 'fee contracting' or 'management fee contracting'. As far as this book is concerned, the definition will be taken from the JCT Management Building Contract (MC) and the NEC3 Option F – Management Contract.

According to the JCT MC, the management building contract is appropriate:

> • *for large-scale projects requiring an early start on site, where the works are designed by or on behalf of the Employer but where it is not possible to prepare full design information before the works commence and where much of the detail design may be of a sophisticated or innovative nature requiring proprietary systems or components designed by specialists;*
> • *where the Employer is to provide the Management Contractor with drawings and a specification; and*
> • *where a Management Contractor is to administer the conditions.*
>
> *The Management Contractor does not carry out any construction work but manages the Project for a fee. The Management Contractor employs Works Contractors to carry out the construction works.*
>   *Can be used:*
> • *where the works are to be carried out in sections;*
> • *by both private and local authority employers.*

Therefore, in a management contract, the subcontracts (or 'work packages' as they are more accurately described) are direct contracts with the contractor. The employer has no contractual link with the work-package contractors. As stated in the NEC3 Guidance Notes for Option F – Management Contracts:

> *If the Employer wishes to be a party to the construction sub-contracts, a management contract is not appropriate. He should instead appoint a construction manager as the Project Manager ...*

Management contracting and fee contracting have a long history of specialist use in the UK dating from the 1920s. It can almost be regarded as a variation of the cost-plus method except that the contractor is more involved in the design process and takes responsibility for the site establishment and other preliminary costs.

The main characteristics of the management contracting system are:

1   The contractor is appointed on a professional basis as an equal member of the design team to provide construction expertise.
2   Reimbursement is on the basis of a lump sum or percentage fee for management services plus the prime cost of construction.

3   The actual construction is carried out by works or package contractors who are employed, co-ordinated and administered by the management contractor.

Many definitions of the method exist, but all contain, in part at least, these fundamental features. The management contractor therefore employs and manages works contractors who carry out the actual construction of the project and the management contractor is reimbursed by means of a fee for their management services and payment of the actual prime cost of the construction.

The system was mainly developed in the USA as a result of public-sector clients requiring phased construction and greater co-operation between consultants and clients. Known in the USA as Professional Construction Management (PCM), it was rather informally used, but as the cost of construction increased during the early 1970s and delayed projects became more frequent, PCM began to be used more frequently as a method of keeping contractor control over costs and time.

In the mid-1970s, the major clients restricted construction management firms to acting solely in a consultant capacity in order to preclude them from undertaking any of the direct works. At the end of the decade, the same powerful organisations further limited the use of the PCM method as a result of difficulties experienced in ensuring that there was enough incentive to perform, problems regarding liability and the need for a firm price tender before commencement on site – the wheel had turned full circle!

Although fee contracting was used as early as 1928 in the UK by Bovis, it was not until 1969 that pure management contracting gained recognition when the Horizon project, a large complex cigarette manufacturing factory for John Player in Nottingham, was built using this method.

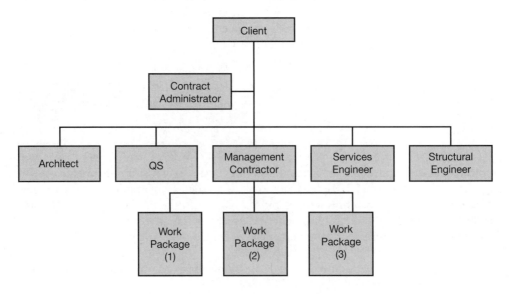

Management Contracting

7.1 Management contracting relationships

Whereas in the USA the growth in PCM was led by architectural and engineering consultancy practices, in the UK the major contracting companies were the first to offer management contracting services to their clients, with the number of contractors increasing dramatically in the late 1970s and 1980s.

Before, and for most of, the 1970s the large majority of, if not all, management contracting projects were in the private sector, and it was not until the end of the decade that the method began slowly to be accepted and used by public clients who had succeeded in overcoming the bureaucratic difficulties associated with public accountability. After initial reluctance, the use of the method increased, mainly as a result of a perceived success rate and good track record, with both local and central government in the UK carrying out major projects using the system.

In the late 1980s and up to the mid-1990s, growth in the use of the management contracting system declined, when measured by both value and number of contracts as can be seen in Tables 3.1 to 3.6 in Chapter 3. Be that as it may, the launch of the JCT Management Contract in 1987 ensured that the method joined the other established and officially approved and accepted procurement systems on offer to the industry's clients.

However, during the UK recession in the early 1990s, a number of influential clients publicly declared their intention not to use the management contracting system on any future projects because of the difficulties they experienced when using this method and the perceived lack of evidence of best value as the contractor was appointed very early in the process. The problems that have led to clients' apparent disenchantment with this method are those of delays caused by trade/package contractors over which the client has little or no control, along with the additional costs incurred when using this system, the lack of a specific contract sum and the difficulty experienced in achieving reasonable quality standards at the beginning of construction.

Over the last three decades, management contracting has become well established in the UK as an acceptable alternative to other, more conventional methods of procuring buildings of all types. However, the RICS 'Contracts in Use' survey for 2004 still found less than 1 per cent of contracts by value to be using this method (see Table 3.6).

A number of reports in the 1980s considered that the trend towards the increased use of management contracting that was seen in the 1980s might not be so marked in the future, and it would appear that this forecast has proved to be correct. The management-contract arrangement does not fit neatly into the conventional pattern of pre-contract and post-contract stages, as the 'contract point' (i.e. the appointment of the contractor) occurs at or close to the beginning of the design stage.

The question of timing of the appointment of the management contractor is crucial, and there are two main ways of doing this. In one approach, the management contractor may be engaged in advance of the architect and other consultants and made responsible for their appointment and the reimbursement of the design team. The more usual arrangement, however, and that laid down in the JCT Practice Note, is for the client to directly appoint the consultants in advance of the engagement of the management contractor and to be responsible for their reimbursement and management.

7.1.2 Management contracting procedures

Where management contracting is the procurement route of choice by the client, the project procedure normally follows three phases:

1   pre-appointment of the management contractor
2   preconstruction period
3   construction period.

Pre-appointment of the management contractor

This period will cover many of the early activities usually associated with the conventional pre-contract stage, i.e. carrying out of a feasibility study, formulation of the brief, etc.; however, they will be carried out in this case on an accelerated timescale. During the early part of this stage, the employer appoints the contract administrator, an architect, a quantity surveyor and any other professional advisers considered necessary – these consultants will form the design team, which initially assists in the preparation of the brief and prepares outline drawings and a performance specification that describe, in general terms, the scope of the project. This is equivalent to stages A to C of the RIBA *Plan of Work*.

Once this outline design and feasibility have been prepared, the employer will invite tenders from management contractors using one of the approved methods of selection and appointment. All of the various methods are likely to contain the following fundamental elements:

1   A longlist is formulated of management contractors, probably of some ten to fifteen in number.
2   These contractors are invited to submit a written proposal to a brief description of the project including the timescale and estimated cost of the works.
3   On the basis of this initial proposal, three to five management contractors are selected to proceed to the next stage and are issued with a questionnaire, the answers to which will provide an accurate profile of their organisation and experience of management contracting. They are also provided with detailed information about the project, its design to date, programme and anticipated cost aims, and invited to submit their final proposals for managing the project and, separately, their fee bid – expressed as a percentage of the total project cost – for implementing these proposals.
4   It is also usual for the contractor to be asked to state the reimbursement that they will require for providing the necessary management services during the preconstruction period.

The evaluation of the contractors should be made on the basis of their written proposals and verbal presentation and should not be determined solely by the level of the individual fee proposals. Great emphasis needs to be placed upon the importance of the way in which the management contractors propose to manage the project and the content and manner of their project team's live presentation of this aspect of the selection process.

The criteria for the selection of the successful management contractor should include:

1   Experience of management contracting and a previous record of success in performing this and the conventional general contractor's role on projects of a similar size, complexity and type and completing work on time and within budget.

2   A detailed understanding of all aspects of the scheme and the ability to match their management team to the needs of the project – this is particularly crucial in the case of the project manager, whose competence and experience will be vital to the success of the project.
3   Adequate financial resources, with sufficient capacity for the bonding and insurance requirements of the project.
4   Sufficient standing within the industry to ensure that any difficulties with suppliers or trade/package contractors are dealt with expeditiously.
5   The ability of both the head-office organisation and the site project team to co-operate with the client and consultants throughout the life of the project.
6   The level of the management fee and payment terms.

As with most other contractor selection procedures, the aim is not necessarily to achieve the lowest fee and other charges but to select the management contractor with the right experience, resources and managerial skills. The sour taste of poor service lingers long after the sweet smell of lowest price has disappeared.

Once the management contractor has been selected, they will enter into a formal contract with the employer – the JCT05 Management Contract or NEC3 Option F Management Contract are both now accepted as the industry standard – and the second phase of the project commences.

Preconstruction period

During this period, the management contractor provides certain services to the client's design team. These will normally include:

1   the preparation of the overall project programme;
2   the preparation of material and component delivery schedules and the identification of those items which require advance ordering;
3   advising on the 'buildability' and practical implications of the proposed design and specifications;
4   the establishment of agreed construction methods, particularly those affecting the design;
5   the preparation of a detailed construction programme;
6   advising on the provision and planning of common services and site facilities;
7   advising on the breakdown of the project into suitable packages for trade and/or works contractors;
8   preparing a list of potential tenderers for the package and pre-qualifying them by investigating their capabilities and financial standing;
9   assisting in the preparation of tender documents and obtaining tenders from approved trade/works contractors and suppliers;
10 the preparation, in consultation with the client's consultants, of the documentation necessary to ensure the efficient placing of the proposed trade/works/suppliers' contracts or orders.

A number of these responsibilities are normally included in the duties of the design team, which, while remaining responsible for the design of the building, must be prepared for a dialogue with the management contractor, for example, on methods of construction and buildability in order to reach an agreed design which best meets the client's needs in the most efficient way possible.

During this period, the management contractor has other functions, which must be carried out before the end of the preconstruction phase, including the agreement of the project cost plan/estimate of prime cost with the cost consultant.

Other matters relate to the agreement of the services to be provided by the management contractor: the terms of the joint policy for insuring the project; the identification of the management contractor's resident site manager; and the agreement of the common site facilities and services that will be provided or secured by the management contractor.

When the contract administrator is satisfied that all the preparatory work has reached a stage when it will be practicable to commence the construction of the project, they will notify the client, who will then make a decision whether the project will proceed or not.

If the project is aborted at this stage, the management contractor and design team are reimbursed their costs to date in accordance with their respective contracts. In the case of the contractor, they will be paid a preconstruction-period fee, which will have been agreed and stated in their contract with the client.

If, however, the management contractor is notified that they are to proceed, they will be given possession of the project site on the contractual date for possession and become responsible for completing of the project by the date for completion or any extended date fixed under the conditions. Once the site is handed over to the contractor, they will also be responsible for insuring the site and works, as is the case under the traditional procurement route.

Construction period

The management contractor does not normally carry out any of the construction work themselves, apart from the site establishment and other preliminaries. As mentioned above, the actual work is divided into a series of separate packages, generally for different trade or functional elements of the building, and the management contractor usually enters into a standard form of work-package contract with each of the works contractors for the implementation of these packages, which should contain back-to-back provisions with the main management contract. The JCT and NEC standard forms of management contract also include standard forms of work-package contract which have these back-to-back provisions and are therefore designed to work together.

The duties of the management contractor during this construction period can be summarised as setting out, managing, organising and supervising the implementation and completion of the project using the services of the works contractors.

The selection of tenderers, the establishment of the conditions of the works contracts, the obtaining and scrutiny of works-contract tenders and recommendations for the acceptance of the most favourable offer are all matters which are dealt with jointly by the client, the appropriate consultants and the management contractor.

Payment is made to the management contractor during the construction period by means of standard progress payments or interim certificates based upon the amount of work so far completed. At the same time, directions are given for the amounts that are to be paid to the individual works contractors. The certificates will also include the reimbursement of an instalment of the construction-period management fee and the cost of the management contractor's directly employed resources. Retention is normally deductible on all elements of the progress payment, including the management contractor's fee.

The monitoring of the cost of the project against the contract cost plan is carried out on a continuous and regular basis by the architect and consultant quantity surveyor, with the assistance of the management contractor, in order that accurate and up-to-date information on committed and future costs is always readily available.

The consecutive receipt of tenders for the works contracts enables the consultants to maintain the overall budget figure and contract cost plan by the redesign of later packages if this is judged to be necessary in the light of the result of earlier tenders. The management contractor's own costs should also be monitored against an agreed figure contained within the contract cost plan.

The architect has the power to issue variations to the scope of work of the project, generally described as 'project changes', and also for the works contracts, described as 'works contracts variations'. However, under the NEC3 Option F, variations to the scope of the works set out in the contract data can only be made via a separate contract, so it is always best to be aware of how your particular contract deals with changes and variations. As with most other procurement systems, variations to the original scope should be avoided if at all possible, but, should such variations be necessary, it is essential that the management contractor and other appropriate consultants assess the effect of the proposed changes on programme and cost before a decision to proceed is made.

The formulation of the construction programme necessary to achieve the completion date is the responsibility of the management contractor, and although extensions of time can be granted under the standard forms of management contract, these may not necessarily affect the actual construction completion date. Although this is quite a technical point, it is good practice to always separate the 'contract period' (i.e. the difference between the date for possession of the site and the date for completion as stated in the contract) and the 'construction period' (i.e. the actual programme and progress). This approach is intended to ensure that the consultants and the management contractor concentrate on completing the project on time rather than on formulating reasons for obtaining an extension of time of the contract period.

During the construction period, the management contractor is responsible for supervising the construction operations and ensuring that all of the work is built to the standards originally defined by the design team. This responsibility can be supplemented by the appointment of clerks of works, with the cost being borne by the employer.

Should any of the works contractors be unable or unwilling to remedy any defects identified in the work they have carried out, it may not be possible to recover the cost of rectifying the faults. In line with the philosophy of the management contractor's status as a consultant, the cost of the necessary rectification usually rests with the client, and clients are therefore advised to establish a contingency fund to cover this risk.

Once practical or substantial completion of the works has occurred and has been certified by the architect or contract administrator, handover of the project is carried out in the normal traditional manner, with the management contractor providing the following:

1   A written undertaking that the project conforms to the original or amended brief.
2   A written undertaking that all costs have been paid, that there are no claims, litigation or the like pending and that all works contractors' accounts have been finalised or will be finalised within a maximum period of twelve months.
3   'As constructed' drawings and a comprehensive specification for all aspects of the project.
4   Commissioning and test certificates, servicing and maintenance manuals, and operating instructions for all elements of the building and services including equipment and plant.
5   A written programme of recommended maintenance for the building, services and plant/equipment.
6   A written undertaking confirming that final testing and balancing of the building services will be carried out once the building is fully occupied and operational.
7   Written confirmation of the date of commencement and completion of the defects correction period.
8   A certificate, issued by the architect, confirming that the project is to their total satisfaction.

It cannot be sufficiently stressed that the management contract is quite a high-risk form of procurement for the client, and it is therefore essential that when using this system they are continuously involved in the project and that the contract administrator or project manager is highly competent and experienced.

7.1.3 Is management contracting better?

This question is addressed in terms of the three criteria of time, cost and quality.

Time

Management contracting comes into its own when short design and construction periods are required or where an early start, and/or early completion, is important to the client. On major projects where these criteria must be met and the use of conventional methods of procurement is not considered appropriate, management-orientated systems have flourished as a result of their perceived ability to fragment the building operations and enable design and construction to overlap and run concurrently.

Many clients do in fact consider that management contracting allows an earlier start to be made on site and also results in speedier completion of the project. Clients also believe that the use of this system results in more reliable predictions being made as to the eventual length of the construction period, as the advice is given by those responsible for the planning of the works.

Various studies have taken place over the years to try to assess the benefits of management contracting against the other forms of procurement. Generally, it has an overwhelming advantage in speeding up the construction process. In addition, the risk of delay is reduced as a result of the system's flexibility and its ability to accommodate and overcome difficulties and changes. The savings in time are achieved by the overlapping of design and construction and from the expedition by the management contractor of the production of the project design. The construction process itself, however, often may not be accelerated as a result of using this method because the design of later packages may not have started, meaning that co-ordination issues could become more difficult.

Cost

The client's perception of the financial consequences of management contracting was that it tended to cost more than conventional methods. Furthermore, the uncertainty regarding the final end cost was seen as a distinct disadvantage.

A survey of ten experienced clients found that:

1 All believed that management contracting was more profitable to the contractor than to the client.
2 Thirty per cent of the clients surveyed were of the opinion that management contracting involved fewer claims, 40 per cent believed that the method involved the same number of claims as other methods and the remainder did not believe that the method involved fewer claims.
3 Only 20 per cent believed management contracting was cheaper than other methods, 40 per cent thought the method produced the same level of cost and the remainder were of the opinion that it was not cheaper (Naoum and Langford 1987).

The use of management contracting may help to minimise costs by improving build-ability (and therefore design), by better packaging of activities to enable competitive tenders for all of the work to be obtained and by the phasing of completion to allow phased letting and occupation. The actual project costs may not necessarily be any lower, although it must be acknowledged that the savings of time that can be achieved may well result in cost savings, or additional income, offsetting the extra construction costs. There is also a suspicion that, relieved from the constraints of competition, the management contractor's expenditure on site establishment, plant and other preliminaries is less than competitive, but by careful tender analysis and benchmarking at the start of the project, this may not be a major hurdle.

Problems that emerged during the late 1980s and 1990s have resulted in management contracting becoming even less rewarding in straight financial terms than other procurement approaches. Trade contractors became reluctant to undertake projects at competitive prices because of the onerous contract conditions imposed upon them by management contractors and because of the high level of risk they were being asked to accept. Extra costs had been incurred as the result of the duplication and overlapping of site management and other services by the principal and subsidiary contractors. This was mainly due to the lack of a proper definition of common services and project supervisory responsibilities by the management contractor.

These issues were temporary phenomena and growing pains of management contracting which have now been dealt with. The use of the JCT and NEC management contracts has had a stabilising effect and has brought the system into the main family of procurement routes.

The absence of a fixed-price contract sum when construction commences is seen by many clients as a distinct disadvantage, and by some as a reason for not using the system at all. Supporters of the method maintain that the lack of such a safeguard does not necessarily mean that there is less control over the eventual construction cost, as strict supervision can be exerted over this aspect of the project. This contention is based upon the fact that each package of work is normally the subject of keen competition between construction contractors, and it is therefore possible to monitor costs closely and, if necessary, adjust later works packages in order to absorb any cost overruns incurred so far, subject, of course, to the client's acceptance of the reduction in scope or specification of the package in question.

The fragmented nature of the construction element of this procurement system allows the final account for the project to be settled progressively over the duration of the works and ensures rapid settlement once the project has been successfully completed.

Quality (functionality)

Provided that the management contractor is appointed during the early stages of the project and in advance of any decisions being taken regarding the detailed design of the building, by advising on buildability, construction methods and techniques, and the economics of the proposed design, they should be able to make a major contribution to achieving an agreed design which fully meets the client's needs and is capable of being efficiently, economically and speedily implemented.

However, for this theoretical advantage to become a reality, it is essential that the design team is prepared to enter into an open dialogue with the management contractor on all aspects of the project and to ensure that they are well briefed in terms of the design constraints and the client's requirements.

The client must therefore ensure that their consultants are fully aware of and committed to the philosophy of the system, as well as being open and receptive to suggestions from the management contractor on any aspect of the design as a full member of the project team seeking solutions to problems that arise during both the design and construction stages of the project. Notwithstanding such co-operation and collaboration, the design team remains fully responsible for the design of the project.

In practice, the amount of assistance that the management contractor can, or is allowed to, give during the design stage is probably limited as a result of being employed at a mid-stage in the design process or their comments and advice not being accepted, even ignored, by the remainder of the design team. To all intents and purposes, this issue has been settled over the past fifteen to twenty years, as contractors now have substantial experience in the design stage and early involvement of the contractor is considered beneficial.

On the question of the quality of the completed project, the general perception appears to be that the quality standards achieved are at least as good as those obtained on conventionally procured projects, and in some cases have been better. It should, however, be noted that whereas quality control is the primary responsibility of the management contractor, the architect may also wish to be represented on site in order to monitor this aspect of the project. In these circumstances, there may be some duplication of site supervision and costs associated with this activity, as well as the possibility of friction between the management contractor and the designer, which could be to the detriment of the client. This potential problem should therefore be resolved before construction commences on site.

Other characteristics

In theory, this system enables the client's consultants and the management contractor to become a professional team of equals committed solely to meeting the client's needs. In addition, the emphasis on the management aspect of construction activities should be beneficial and should ensure the effective control of the project.

Specifically, the following characteristics are associated with this method:

1   A high degree of flexibility is built into this method, which enables delays to be overcome or reduced, changes to be absorbed and rescheduling of works packages to be implemented.
2   The fragmented structure of the construction process, i.e. the use of works packages, means that the financial failure of any of the works contractors has only a limited effect on the total project.
3   The management contractor's knowledge and experience should ensure that industrial relations on the project are better than when using more conventional procurement methods.
4   Research and development of new techniques and processes may result from the greater involvement of the contractor during the design stage.
5   The initial philosophy, which has been fundamental to this system since its inception, of the management contractor's liabilities being limited, like all other members of the professional team, to any negligence in the performance of their management function means that the client must accept far more risk and responsibility than when using many other procurement systems.
6   Contrary to the original principle, management contractors have been made by some clients to accept the risks associated with construction rather than just the management of the project. Specifically, clients have made management contractors responsible for construction contractors' time overruns, remedying of work-package defects and even some design responsibility.

These responsibilities bring risks which are similar to those associated with design and build arrangements and place the management contractor in a position where they need to decide whether to accept the unreasonable risks themselves, and pass on only the reasonable hazards to the works contractors, or whether to ensure that the latter takes all of the risk with the inevitable consequences. The wide use of the JCT form of management contract may have ensured that this deviation from the original philosophy has not irreparably damaged the method.

There is little doubt that this procurement system increases the amount of administrative effort required of all of the participants, but particularly the client, and produces a much greater amount of paperwork than many of the other methods.

All of the individual methods of procurement contained within the management-orientated category have a number of characteristics in common, and, thus, it is not proposed to reach any conclusions as to the advantages and disadvantages of the management-contracting system until all of the members of this group have been considered.

It is, however, fair to say at this stage, particularly in light of the tribulations that this particular system has experienced in the past, that management contracting is not, as it has often been seen, a panacea for all of the construction industry's ills. Rather, if properly applied, it is a viable alternative to conventional methods, especially where the project is large and/or complex, where there is a need for flexibility and where there is a need for an early start or rapid completion.

7.2 Construction management

7.2.1 Definition and history of construction management

In the construction management procurement method, the management service is provided by a fee-based consultant, a specialist construction manager or a contractor and all construction work-package contracts are directly agreed between the client and the individual trade (package) contractors. Therefore, the fundamental difference between this procurement system and management contracting is that, with this approach, the client themselves enters into a direct contract with the individual package contractors, whereas with management contracting, the contracts are between the management contractor and the work-package contractors. The construction manager then acts as a co-ordinator between all the separate packages. The main characteristics of the system are:

1   The construction manager is appointed as a consultant during the initial stages of the project and has equal status with the members of the design team.
2   Reimbursement is made by means of a lump sum or percentage fee for management services.
3   The physical construction of the project is carried out by work-package contractors who are employed directly by the client and co-ordinated, supervised and administered by the construction manager.

The JCT05 Construction Management (CM) standard form of contract states that the contract is appropriate:

> - *where a Construction Manager is to manage the project on behalf of the Client; and*
> - *where the Client is to enter into direct separate trade contracts using the Construction Management Trade Contract (CM/TC) or a special Trade Contract.*
>
> *and can be used:*
> - *where the works are to be carried out in sections.*

In the USA, architects and engineers began to offer construction management services to their clients during the 1960s. This method has usually been referred to in North America as construction project management (CPM).

It was not until a decade later, during the 1970s, that this method began to be used in the UK in response to demands from clients of the industry for greater certainty about the overall performance of their project, and in particular the need to control the risks of cost and time overruns and detrimental external influences on very large and complex projects.

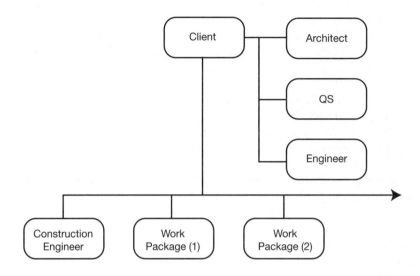

7.2 Construction management contractual relationships

These problems were further exacerbated when, in the mid-1980s, a depressed and uncertain world economy, coupled with high interest rates, exposed commercial developers to large financial risks. In these circumstances construction management appeared to provide extra management expertise and a control structure which was seen as being necessary to deal with the increased levels of uncertainty arising from these effects.

Since that time, the system has been increasingly used by major developers, particularly on very large projects in London and other major cities with apparent success, although often to the detriment of UK consultants and contractors, who seem unable to meet the standard of service and cost parameters demanded of, and achieved by, American and other European design and construction organisations.

The use of construction management in the UK by the public sector during its early years was limited, but after a period of experimentation government agencies have used this procurement system on appropriate projects and it has now mostly overtaken management contracting in terms of its use. See RICS 'Contracts in Use' data in Chapter 3, sections 3.3 and 3.4.

7.2.2 Construction management procedures

There is considerable variation in construction management procedures and practice, and therefore the following discussion concentrates on that process that has become accepted within the industry as good practice.

The contractual relationships in the system are shown in Figure 7.2. The construction manager should be appointed as early as possible in the life of the project, alongside and of equal status to the design team, with both parties dealing directly with the client. During the preconstruction phase, the construction manager will carry out the same roles as the management contractor described in section 7.1.2. The construction manager is directly responsible to the client for co-ordinating and managing the activities of the members of the design team and will also co-ordinate with the client's own in-house project management personnel, so that the client organisation is therefore more intimately involved with the management of the project than if a more conventional procurement method had been used. As this form of procurement is normally used only for very large and complex projects, the clients are almost always large corporations with extensive construction programmes and will therefore have their own directly employed professional staff. They are experienced clients, as defined in Chapter 2.

There are five major stages within the construction management process, which are broadly similar to the stages in the RIBA *Plan of Work*:

1   concept
2   detailed feasibility
3   scheme design
4   design completion and construction
5   completion.

Concept

During this stage, an initial project brief, including cost, quality/functionality and time parameters, is prepared, sketch designs are produced and a feasibility study is prepared by the designer and construction manager, who should preferably be appointed during this phase. Such appointments can be limited in time and scope in order to protect the client against the consequences of the project not proving feasible.

Elaborate procedures were developed for the selection of the designer and construction manager during the initial use of the system in the UK, this unusual thoroughness probably reflecting the comparative novelty of the use of the approach at that time. As the use of the system has developed and standard forms of construction management agreement have become available, the selection methods have been simplified and are now similar to those used for the selection of management contractors, with the eventually shortlisted organisations making formal submissions, consisting of an offer to provide a specified service for a specific fee, on the basis of more detailed information about the project provided by the client.

Once formal submissions are received, each firm is interviewed in order to discuss the submission and allow the client to meet the designer/construction manager's key personnel. The normal form of interview is a presentation by the firm followed by questioning by the client in order to identify weaknesses, obtain missing information and evaluate the organisation's resources, personnel and capabilities.

Finally, the choice of the designer/construction manager is made by the client based on a full discussion and evaluation with their advisers of the shortlisted organisations' strengths and weaknesses against a list of weighted criteria. The construction manager may also fulfil the role of CDM co-ordinator, who is required to be appointed before any design work is undertaken.

The designer's duties will include all of those responsibilities normally associated with building projects, i.e. the preparation of all stages of the design, obtaining approvals from statutory authorities and other third parties, preparing cost plans, etc. In addition, they will be responsible for co-ordinating the work of other consultants and advising on their appointment, checking and approving any design work carried out by the works contractors, contributing to the preparation of the project brief and generally working closely with the client and construction manager on all aspects of the project.

Once appointed, the construction manager's responsibilities, all of which will need to some extent to be carried out in co-operation with the client and the designer, will generally consist of:

1   The overall planning and management of the total project from their appointment to completion.
2   The assessment of the design to ensure efficiency and buildability, compliance with budget costs and programme, value for money, economy and design-stage health and safety issues.
3   The provision of advice on real construction costs and the establishment of cost budgets for the project and the works packages.
4   Identification of all statutory requirements, value engineering including advising the client of design cost implications and suggesting alternative methods, if necessary, of meeting these requirements.
5   Responsibility for health and safety issues for the site during the construction stage.
6   The planning, management and execution of the construction phase of the project, including dividing the work into appropriate packages, obtaining tenders for each work package and ensuring that pre-ordering of essential materials or equipment is undertaken at the appropriate time.
7   During the currency of the project, ensuring that the client is kept informed on progress, costs and levels of future expenditure; evaluating and issuing variations once they have received the client's approval; agreeing interim payments and final accounts with the works contractors; and advising the client on actual or potential contractual claims.

These duties must be incorporated into some form of legal agreement which will establish a contract between the client and the construction manager and determine the roles, responsibilities and liabilities of each of the parties.

In the UK, the JCT has issued standard forms of contract within the 2005 suite of contracts for construction management, including the following documents:

a   *Construction Management Appointment (CM/A)* – for the firm providing the construction management service;
b   *Construction Management Trade Contract (CM/TC)* – between the client and each trade contractor/work package. This contract can only be used in conjunction with CM/A above;
c   *Construction Management Guide (CM/G)* – giving guidance on how to complete the contract, advising on any unfamiliar terms and specifying who should complete it;
d   standard forms of collateral warranty given by the construction manager or trade contractor to a funder or purchaser/tenant.

A collateral warranty is a means by which a third party with an interest in a project can set up a contractual link with the actual builder of the project, so if any defects appear they have redress under the law of contract. This followed a legal case in the UK which held that unless the defect caused physical injury (e.g. a wall fell down and hurt somebody), any 'economic' loss (e.g. rising damp causing staining on the walls and reducing the rental value) could not be recovered under the law of tort or negligence. This is obviously a very brief explanation of a complex legal issue which is explained further in contract law texts.

In terms of payment to the construction manager during the design stage, the client undertakes to pay a preconstruction-period management fee, which can change if the circumstances of the project change.

Detailed feasibility

During this stage, the designer develops the sketch designs and, together with the client and construction manager, enlarges and firms up the project brief, the cost plan and the means by which the project will be managed and controlled in order to determine finally whether the project is still viable. As a result of this exercise, the client decides whether the project should proceed to the next stage.

Scheme design

The scheme design is finalised during this stage once it has been determined that it satisfies the project brief and can be implemented within the cost plan. If this is the case, approval to proceed is given by the client.

The construction manager will primarily be involved in the preparation of the project cost plan and with forecasting the client's cash-flow requirements. They will also be defining the works packages, preparing tender documents, selecting potential works contractors, establishing the project programme, construction method statements, health and safety documents and finalising management procedures.

This is the critical stage in terms of determining whether the client is in a position to authorise the commencement of completion of the design and in terms of obtaining of the first bids to allow work to commence on site. Such authorisation can only be given once assurances are received from the designer and construction manager that the client's brief has been, and will be, fulfilled.

Design completion and construction

During the early part of this stage, the design is completed and the construction manager begins to implement the tendering procedures by obtaining, reviewing and evaluating bids from works contractors and, together with the designer, making recommendations to the client for acceptance of tenders.

As with management contracting, the construction manager does not carry out any of the construction works themselves, but they are responsible for the control and co-ordination of package contractors who actually execute the works under a series of direct work-package contracts with the client. The construction manager acts under this arrangement as the agent of the employer. As mentioned above, the works contracts should be based on the standard forms of contract designed for the purpose, but can also be in-house forms of the particular construction management firm. If this is the case, it is important for the client to satisfy themselves that the package contracts have 'back-to-back' provisions with the construction manager's contract so that there is minimum scope for ambiguity between the contracts.

During the construction period, the main tasks of the construction manager are the controlling of the cost of the project against the agreed budget, estimating the cost of design and construction proposals, using value-engineering techniques to review design proposals, monitoring tender costs and adjusting the content of future works packages to ensure adherence to the approved estimate of the cost of the work.

Time management, or programme control, is another fundamental task for which the construction manager has responsibility. Its purpose, once a programme for the project has been established, is to ensure that the work is completed within the time agreed with the client by monitoring progress of the individual work packages and correcting any deviations from the established programme.

Apart from completing the design, and dealing along with the construction manager with any necessary variations to the original scheme, the designer's main task is to examine and co-ordinate the designs submitted by other consultants and the works contractors, ensuring that they comply with the original specifications and the project brief and, if satisfactory, sanctioning their implementation.

As the designer is not always the best person to ensure the quality of the work on site, the construction manager should take responsibility for overall quality control. The whole project team must be involved in establishing the initial standards for works contractors and subsequently monitoring quality and ensuring that defective work is remedied.

Interim applications for progress payments submitted by the works contractors are examined and assessed by the construction manager, together with the designer and other appropriate members of the design team, and recommendations made to the client for payment depending on the contractual terms of payment.

The validity on any claims for variations, delays, etc. submitted by the works contractors is established by the construction manager, who assesses and agrees their value and recommends the necessary action to the client after discussion with the appropriate designer. Final accounts are dealt with in the conventional manner.

Completion

The final elements of the construction process, i.e. the inspection, testing, commissioning and handover of the separate works packages once substantial completion has been effected, the issue of individual completion certificates to works contractors and the monitoring of defects liability/correction periods, are the responsibility of the construction manager working in liaison with the other members of the professional team. Clearly, this is a complicated process, as each individual work package will have its own completion, snagging list, etc. The defects liability/correction period for work packages is often linked with the appropriate period for the main contract, so whereas the period is often given as twelve months from completion, this is the completion of the whole project, so for early trade packages, the defects period could run into years.

7.2.3   Is construction management better?

Again, the three criteria of time and cost and quality are now applied in the context of the performance of the construction management procurement system.

Time

The construction management method has two basic characteristics which should lead to faster completion.

First, construction is undertaken by separate works contractors with each work package being capable of beginning as soon as firm designs are available for that package and the successful tenderer appointed. Therefore, project design and construction can be successfully overlapped. This approach can be reinforced by the construction manager pre-ordering and procuring critical materials and components ahead of construction, and by the fact that construction managers, who are normally responsible for the time management, programming and monitoring of the design activities, can ensure that information delays are minimised.

Second, the use of this system should ensure that designs are more easily built as a result of the construction manager's input at design stage and the consequent avoidance of unnecessarily complicated work and the minimising of temporary work and falsework.

A number of government-sponsored and commercially sponsored reports, when comparing conventional methods of procurement with less conventional approaches such as construction management, confirmed that, on average, the design and construction phases of projects using the less conventional route were completed more quickly than when using conventional procurement systems, and generally advise the use of methods such as this when early completion of the project is an important or crucial factor.

Many contemporary reports on construction management projects show that those involved with the project believe that the enlightened management approach being adopted on site and the fact that a proactive attitude to difficulties and problems has been engendered has led to the dramatic improvement in speed of construction over conventionally managed projects.

Cost

The use of works packages ensures that competition is still achieved on a major part of any project being carried out using construction management. The fact that the employer enters into direct contracts with the work-package contractors ensures a greater measure of control over costs and the overall financial state of the project. It is also possible, during the construction phase of the project, to adjust the scope or specification, and thus the cost of the uncommitted work should the contract already awarded have exceeded its estimated cost.

Research into construction management projects shows that they have a lower average cost than the management contracts that were surveyed, and that the average cost overrun for the former method is lower than the corresponding figure for the latter. However, very little further specific evidence appears to exist with regard to the comparative cost of this method, and it must therefore be a matter of conjecture as to its relationship with other procurement methods in this respect.

Although the question of the level of direct construction costs remains unresolved, there is a real possibility that the effect of the likely saving in time that will be achieved when using this method will mean that in terms of the client's total real costs a net saving may be achieved as a result of reduced financing changes, earlier rental income, etc.

The system's association with the fast-track approach, which is demanding of management skills and can lead to heavy expenditure on resources in order to achieve the necessary extra speed of construction, makes for extreme difficulty in reducing direct costs when using this method. In practice, such cost benefits as can be achieved depend upon more mundane day-to-day management activities such as the close scrutiny of the works contractor's accounts, contra-charges or any other aspects of normal commercial contracting.

One of construction management's main weaknesses is that it leaves the client vulnerable when, at the commencement of the project, they have entered into the first works contract and they are therefore irrevocably committed to the project without knowing the final out-turn costs. Unfortunately, this is the main disadvantage for all early commencement procurement routes and has to be accepted by the client if they consider early start an important criterion. For this reason, it is crucial that accurate budgeting and estimating is carried out throughout the project to give the client regular and accurate cost reports.

There is, of course, no lack of evidence that on many projects various forms of incentives, designed to increase the construction manager's risk, have been implemented in an effort to overcome this weakness and to ensure that the estimated final cost is not exceeded. The main way of doing this is to ensure that the contractor gives a guaranteed maximum price (GMP) on the understanding that the client will not pay any more than this amount for the total project. The client may also offer a target contract, where any savings or increases on the target cost are shared between client and contractor in a pre-agreed ratio (painshare or gainshare). In these cases, it is likely that the careful, detailed planning and costing of the project, which is normally characteristic of this method, will be carried out with even more caution, with the result that the usually high preconstruction costs associated with this method of procurement may be further increased.

There is thus very little evidence to suggest that the use of construction management will result in large cost savings, or the converse, and it therefore appears that the best that can be said is that this system, when correctly used, ought not to perform worse than conventional methods and on large, complex and innovative projects is likely to produce better results in terms of cost than when using many other procurement methods.

Quality (functionality)

There is little evidence to support the view that there is an improvement in the quality of the project's detailed design as a result of the involvement of the construction manager during the initial stages of the development. Many traditional designers may suggest that contractors have a poor understanding of the design process and give it too much of a commercial bias, being unable to adopt an objective approach to design matters. However, design and build and early contractor involvement are now a normal and readily accepted procedure, and the experience gained by contractors in the design process has generally improved the product of the industry.

Provided that the construction manager implements the correct quality-control procedures during the construction stage to ensure that works contractors are aware of the standards expected of them, and that their performance is strictly monitored, a high level of quality can be achieved and, if the manufacturing industry experience is repeated, will be accompanied by an increase in productivity.

Overall, therefore, the combination of an experienced design-orientated construction manager and the correct construction-quality procedures provides a good chance of ensuring better-than-average performance in this area of the project's implementation.

General

The emphasis on management which is inherent in this approach generally results in clients and designers making timely decisions to match the needs of construction and, should the need for design changes arise, being in a position where a concerted effort by the whole project team can be made to minimise the time and cost penalties that could be incurred by the client.

As in the case of proposed late changes in the design, many of the benefits arising from the use of construction management centre on the project team's knowledge and experience working unambiguously in the client's interest throughout the life of the project and leading to improved buildability, the minimising of problems during the construction period and the close matching of the contents of the works packages to the capabilities of the works contractors.

The method also provides for strict control of the design process, particularly in terms of the co-ordination of the production of design information with the requirements of the construction programme. Anecdotal evidence also indicates that a good sense of teamwork and a positive approach to problem-solving is engendered when the construction management system is used.

The last characteristic may be the result of the greater client involvement that the use of this system demands and the better working relationships that result from this, including those between the construction manager and the works contractors that stem from the client entering into direct contract with each individual contracting organisation.

The fact that under this arrangement the client is responsible for direct payment of the works contractors' accounts usually results in an improvement in the cash-flow position of the individual contractors, as they are not reliant on payment via the 'main contractor', with the delays that invariably causes.

Clients should be aware, however, that the use of construction management involves them in additional administrative duties and responsibilities to those accepted when using most other methods of procurement and also increases the risks they carry, particularly those associated with cost overruns, delays and claims. Therefore the client may also need to appoint an employer's agent or client representative to look after their affairs.

Despite the advantages gained by works contractors when involved in construction management projects, it is possible that they can feel more exposed when tied to a direct contract with the client and lack the protection of a main contractor. More of the responsibility for co-ordination and performance rests with them, as does much of the site management and 'attendance' tasks previously undertaken by the main contractor on more conventionally managed projects. Under certain circumstances they can also be allocated onerous design responsibilities for which they may lack expertise, insurance and payment.

As is the case with all building procurement systems, the circumstances of each project and the competence and experience of the design and construction teams need to be weighed against the advantages and disadvantages that have been identified as characteristic of construction management before a decision can be made as to its suitability for use on a specific project.

However, this method needs to be implemented with great care and precision in order to ensure that the potential benefits, particularly speed of construction, can be achieved. On the basis of its track record, construction management if used correctly and efficiently has become firmly established as one of the most effective methods of managing large, complex building projects for experienced clients.

7.3 Summary and tutorial questions

7.3.1 Summary

Over the last thirty years, both management contracting and construction management, as alternative procurement methods, have flattered to deceive. A look at the RICS 'Contracts in Use' survey in Tables 3.1 to 3.6 (Chapter 3) shows that neither route has achieved more than 10 per cent of use, apart from a period in the late 1980s when management contracting was in vogue and was used in up to 15 per cent of contracts by value, but less than 2 per cent of contracts by number – i.e. some very large projects.

Both are, however, useful tools for large contractors to have at their disposal, but it would be unwise to rely on them for anything other than a small amount of turnover. Management contracting is about programming, co-ordination, planning and control. The management contractor does none of the construction work themselves and therefore will not own any major items of plant and equipment. This was seen as a benefit in the 1980s, when business ratios such as return on capital employed were not seen as important, but for various reasons, this has changed and having a strong capital base is now seen as advantageous. This is equally valid with construction management as a procurement route, with the additional point that many consultancy organisations can act as the construction manager, since the work package contracts are held directly by the client.

The major difference between management contracting and construction management is who holds the contract with the package contractors. In management contracting it is the management contractor and in construction management it is the client.

7.3.2 Tutorial questions

1   Compare the types of projects which are suitable for management contracting and construction management.
2   How would a management contractor be appointed for a major commercial office development?
3   Compare the stages of a project under the two systems.
4   As a client for a commercial office development, which system would you prefer, having read this chapter? Give your reasons.
5   Give the potential reasons for the limited take-up of these systems since 1990.

8 Project-specific procurement
Integrated roles

This category of procurement incorporates all of those methods of managing the design and construction of a project where these two basic elements are integrated and become the responsibility of one organisation, usually a building contractor. Other elements, such as funding of the scheme and operation and maintenance when the building is in use, may also be incorporated within the system in the case of PFI and PPP projects.

The *design and build* procurement system is the main member of this category, with variants of that method making up the remainder of the group. The principal variants that are dealt with here are *novated design and build*, the *package deal*, *develop and construct*, *design and manage* and *turnkey* methods of procurement.

8.1 Definition

The term 'package deal' has been used in the building industry for many years as an all-embracing description covering design and build, the 'all-in' service, develop and construct and turnkey contracting.

This imprecise terminology has led in the past to considerable confusion among both practitioners and clients of the industry alike, resulting in misunderstandings as to the nature, characteristics, advantages and disadvantages of the basic design and build system and, in some extreme cases, to its rejection as a possible candidate for use on some projects purely on the basis of its supposedly poor reputation for satisfying the client's needs.

However, for at least the past three decades, although some confusion still exists among inexperienced clients, the term 'design and build' has been almost unanimously interpreted and defined as being an arrangement where one contracting organisation takes sole responsibility, normally on a lump-sum fixed-price basis, for the bespoke design and construction of a client's project up to it practical/substantial completion.

This definition contains three elements which are fundamental characteristics of this system:

a   the responsibility for design and construction lies with one organisation;
b   reimbursement is generally by means of a fixed-price lump sum;
c   the project is designed and built specifically to meet the needs of the client as defined in their initial employer's requirements and developed by the contractor's proposals.

It could be argued that the design and build method is probably the oldest proven procurement system still in use in the UK, as until the middle of the eighteenth century first the client, then the architect and finally the master builder was, in turn, solely responsible for both the design and construction aspects of buildings. Indeed Sir Christopher Wren was more of a design build contractor for St Paul's Cathedral in London, although most people regard him solely as an architect.

At about this time, following the Great Fire of London, the complete separation of design and construction finally occurred, and the approach that we know of today as the traditional procurement system (single-stage lump-sum competitive tendering – see Chapter 6, section 6.1) became the main method used to implement projects.

As this latter method maintained its dominance over all the other systems until the 1970s, it is not surprising that design and build only began to emerge from its period of dormancy after the Second World War, and even then initially only to answer the needs of the ambitious targets set by government for the public housing sector, where most of the buildings could be of a standard design.

Without contractor-designed housing systems, the high housing output figures of the post-war years would not have been achieved, and the credibility and viability of a procurement procedure whereby the contractor acts as both designer and constructor would probably not have been established so quickly.

In parallel with the re-emergence of the design and build principle as a tool to meet local authority housing targets in the UK, greater use of this system for industrial and commercial projects was being made in the USA and, gradually, following the lead from across the Atlantic, private-sector clients in the UK began to adopt the integrated approach being marketed by contractors.

The overheating of some sections of the national economy during the various boom periods resulted in heavy demands upon the construction industry. At the same time, a shortage of construction resources, coupled with claims by contractors of greater efficiency and lower cost when using this method, led to the increased use of contractor-designed 'system' building by the public sector in both the residential and commercial sectors of the market.

Although the use of contractor-designed systems declined in the late 1960s and early 1970s, contractors had by then amply demonstrated their ability, particularly in domestic building, to manage large integrated projects and achieve savings in time. However, little evidence exists that the direct cost savings forecast were ever achieved, probably thanks to the difficulty in comparisons across projects at different times and locations.

However, the state of the national and international economies again produced an opportunity for contractors to satisfy the growth in interest among clients in finding alternative ways of procuring projects that occurred as a result of the 1973–4 oil crisis, when a dramatic increase in borrowing and inflation rates emphasised the need to ensure that projects were both speedily commenced and completed in order to minimise the associated financial risk of development.

At the same time, client dissatisfaction with the performance of conventional methods of building procurement meant that any method where there was single-point responsibility, where the integration of design and construction could lead to savings in time and where fixed-price lump-sum tenders could be obtained was extremely attractive. These heavily contractor-marketed characteristics ensured the growth in use of the design and build system and in turn produced one of the most significant trends in construction procurement in recent years.

Let us now look at the three major forms of integrated procurement.

- *Pure design and build.* In this form, the contractor strives for a complete and self-contained unit where all the necessary design and construction expertise resides within one organisation that has sufficient resources to complete any task that arises. Such firms rarely undertake anything other than design and build contracts and usually operate within a particular region or, more likely, a number of discrete market sectors. In such organisations, all aspects of design and construction have the capacity to be highly integrated, and there is a wealth of experience and site feedback which can be utilised for the management of future projects.
- *Integrated design and build.* In this form, a core of designers and project managers exists within the organisation, but this type of contractor is prepared to bring in design expertise from external organisations whenever necessary. The design and construction teams may well be separate entities within a group, and both design and build and conventional tendered work may be undertaken. Although more effort is needed to integrate the internal and external members of the design and build team, in-house project managers are employed to co-ordinate these functions.
- *Fragmented design and build.* Many contractors, both large and small, and including national builders, operate a fragmented approach to design and build projects, whereby external design consultants are appointed and co-ordinated by in-house project managers whose other main task is to take and refine client briefs into workable contractor's proposals. Under this regime, many of the integration and co-ordination problems of the traditional approach are likely to manifest themselves along with some role ambiguity among the professions as they come to terms with the builder as leader of the design and construction team. The majority of medium and small design build projects are undertaken by contractors from this category.

Clients and their advisers need to be aware of these categories, and their individual characteristics, when selecting contractors who will be asked to submit proposals. Problems can and often do arise when the chosen organisation is not compatible with the client's operational needs and management structure.

It is likely that contractor's costs, and therefore tender prices, will increase as the choice of contractor moves from fragmented design build through the integrated method to the pure form. This increase in costs results from the reduction in the amount of design work subcontracted to external consultants when using the integrated and pure methods and the fact that the 'no job/no fee' arrangement prevalent in the fragmented and, to a lesser extent, integrated form does not apply to the pure arrangement. On the other hand, the comparatively small number of pure design and build contractors operating in the UK have amply demonstrated their ability on numerous occasions successfully to manage large, complex projects which have fully met their client's needs and objectives, and presumably cost less in the long term despite incurring short-term additional costs at the design and construction stage.

Having decided upon the type of contractor who would be most suitable for the project, it is necessary to ensure that potential tenderers have the appropriate kind of design and build experience relevant to the current project and the necessary financial capacity and the other required resources. This is normally done by means of a pre-qualification questionnaire (PQQ) and interview which can be both very detailed and expensive to prepare. Other standard selection procedures, such as obtaining references from previous clients and visits to completed or current projects, are all used as further means of, or aids to, selecting the most appropriate organisations.

The main methods of obtaining tenders for design and build projects are therefore limited to single- and two-stage tendering and negotiation. The actual choice of procedure will depend mainly on the extent to which the employer wishes to negotiate after the submission of the tender, as well as the complexity of the project, although if an element of competition is required the option is obviously restricted to the single- and two-stage methods.

The single-stage procedure requires the selected contractors, usually three or four in number (for design and build projects), to make only one submission to the client, with the preferred proposal forming the basis for the design and build contract with detailed design work, and the obtaining of the necessary approvals, commencing immediately the contract has been awarded.

With the two-stage method, up to six contractors can be invited to submit preliminary proposals in the form of outline designs and budget costs. The favoured scheme is then selected, the design is developed to an advanced stage and the budget cost converted into a firm bid, which, if acceptable, forms the basis of a design and build contract, with work being implemented once the contract has been awarded.

The increased use of resources at bid preparation stage by design and build contractors when compared with other forms of tendering can be a serious problem, and it has often been recommended that reimbursement of second-stage design costs should be made to the tenderers if the project is abandoned by the client. This recommendation does not apply to the negotiated alternative, as this method generally restricts the negotiation to one contractor, with the design and contract sum evolving during discussion between the two parties until the client is satisfied that both elements of the proposal meet their requirements and that the cost has been kept within the budget.

It is necessary when inviting tenders to advise the bidders which form of contract will be used once the most appropriate submission has been accepted. The main standard forms of contract available for design and build are:

1   Joint Contracts Tribunal (JCT) Standard Form of Design Build Contract (DB) 2005;
2   the New Engineering Contract, 3rd edition (NEC3). The extent of the project to be designed by the contractor is given in the Works Information;
3   Government Contract (GC)/Works/1: Edition Three, Single Stage Design and Build Version, 1996;
4   Institution of Civil Engineers (ICE) Design and Construct Form of Contract, 1992;
5   FIDIC Conditions of Contract for Plant and Design-Build (Yellow Book). This contract can be used for both electrical and mechanical plant and for building work designed by the contractor.
6   FIDIC Conditions of Contract for Design Build Operate (Gold Book). Introduced in 2008.
7   client-drafted forms;
8   contractor-drafted forms.

As mentioned previously, the use of nationally agreed standard forms of contract provides a stable framework within which the client and the building team can operate with the minimum of difficulty and confrontation, as each party should understand the terms of the contract and how they operate. The use of bespoke forms of contract or the amendment by one party of the standard form of contract may create ambiguities and uncertainties, leading to disputes. It is worth reiterating the point made earlier that standard forms of contract which have been amended by one party are naturally only going to be amended in favour of that party, consequently adding even more risk for the other party.

8.2 Share of the market

Efforts to identify accurately how much of the annual non-domestic building workload is carried out using the design and build system are made difficult by the probable use of incorrect terminology in such surveys as are available and the resulting difficulty in determining whether the results refer to the main system or its variants.

Of more importance, however, if one is trying to establish which category of client uses which system, is the fact that different clients use the same procurement system to a greater or lesser extent depending on the nature of their business and in some cases the type of project they are implementing at the time. Generally speaking, industrial clients carry out more of their building projects using the design and build system than commercial clients, although the relative share can fluctuate over time.

The use of design and build has increased substantially over the last twenty years, as can be seen from the RICS 'Contracts in Use' survey results (see Chapter 3, Tables 3.1 to 3.6), which show that the value of contracts using the system had increased from 3.6 per cent (8 per cent) in 1985 to 13.3 per cent (43.2 per cent) in 2004. The difference between the two figures in 2004 means that only 13.3 per cent of the total number of contracts used the design and build system, but they represented 43.2 per cent of the total value of work. Therefore the projects using design and build were the larger projects.

Design and build is clearly the fastest-growing procurement system in the UK, and although the level of use of the system can vary considerably depending on the category of client and the type of project commissioned, the share of the market will always be substantial. Having increased dramatically over recent years, it is likely to continue to do so for the foreseeable future, although perhaps at a much slower rate than previously on account of the longer-term effects of economic recession from 2008–10.

8.3 The process

With the contracting organisation taking sole responsibility for both design and construction, the contractual and functional relationships among the employer, consultants and contractor are simplified when compared with most other procurement methods, with communications being reduced, in theory at least, to a single channel. Figure 8.1 illustrates the relationships between the various members of the project team.

It follows that once the decision has been taken to adopt this method of procurement the process should be a simple one consisting of:

- the preparation of the employer's requirements;
- the obtaining of tenders, which includes contractors' proposals (for design, cost and time);
- the evaluation of the submissions on the basis of design, specification, cost and programme;

Design and Build

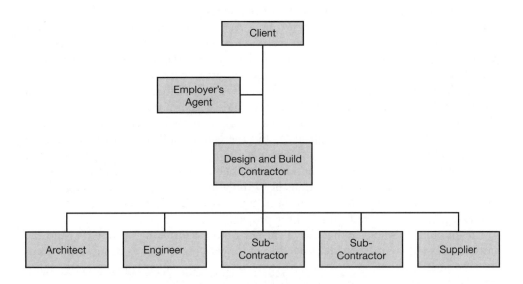

8.1 Contractual relationships in a design and build project

- the acceptance of the most appropriate tender;
- the implementation and completion of the project.

This process is set out in the Code of Procedure for Design Build Tenders issued with most of the standard forms of contract.

This simplicity, however, is deceptive, and a detailed examination of the process reveals the fact that the various constituent parts are, in themselves, relatively complex and contain a number of potential pitfalls for the unwary and/or inexperienced client.

8.3.1 Employer's requirements

The purpose of the employer's requirements is to provide the tendering contractors with sufficient information, in the form of a brief, to enable them to develop their proposals and bids without unnecessary difficulty. The brief therefore needs to include the following information as a minimum:

- details of the supervisory and/or cost consultants;
- a clear statement to the effect that the contractor is required to manage, produce detailed designs and construct the building project to meet the client's specific needs;
- details of the contractor's direct management and labour force;
- details of any subcontractors the contractor intends to use;
- details of the intended contractual relationships, functional relationships and any optional appointment by the client.

The brief needs to be clearly presented and sufficiently comprehensive to leave the tenderers with no doubt as to the precise wishes of the employer while, at the same time, giving them the freedom to make use of their design, technical and managerial expertise as well as their own particular available resources.

In order to achieve this objective, the employer's requirements should theoretically be drafted only in terms of the project's required performance criteria. However, in practice, in order to achieve clarity and avoid misunderstandings the client's needs are more likely to be stated in both performance and prescriptive terms.

This should enable tenderers to be sufficiently aware of the client's requirements in order to ensure the submission of best value bids and easily comparable proposals. However, as a result of commercial pressures to start the project, clients are often unable to provide a clear enough brief, and the lack of sufficient information as to their requirements is an inhibiting factor in arriving at a fair and valid assessment of the contractors' bids and the client's ability to make a choice between them. Clients, however, have a wide range of options in formulating the brief, with anything from a short description of the accommodation required to a full scheme design being appropriate, depending upon the nature of the project.

The amount of influence that the client has on the contractor's proposals, and the associated bid, therefore depends upon the employer's requirements being comprehensive and unambiguous, as well as the extent to which the client is prepared to elaborate on and clarify their requirements before tenders are submitted.

In accepting the responsibility for interpreting the client's requirements correctly, and bidding on a lump-sum basis, the contractor is assuming a much higher risk than when they are provided with more comprehensive documentation in tendering for fully designed projects using other methods of procurement. The burden is increased if the client's needs are not adequately identified and described, and bids may well then be increased to include a premium, in the form of an undisclosed contingency sum, clearly resulting in unnecessarily inflated tenders.

As with all briefing documents, achieving the correct composition of the employer's requirements is critical to obtaining the most advantageous design proposals and bids and the overall success of the project.

8.3.2 Contractor's proposals and their evaluation

The evaluation of design and build proposals and tenders has already been identified as an area where problems are often experienced because of the difficulty of communicating the client's detailed requirements by means of a single written brief, i.e. without the assistance of drawings or a bill of quantities, which can lead to different interpretations for the tenderers.

One way of solving this difficulty is for the client, after the initial appraisal has been made, to interview the design and management team of each tenderer in order to discuss the individual schemes and therefore obtain a better understanding of each proposal. Some clients see this approach as being unethical and contrary to the guidance given in the various codes of procedure on the grounds that discussions with individual tenderers might result in breaches of confidentiality, and believe that such discussions should be limited to the lowest tenderer(s). However, nowadays, the interview is considered by most clients and advisers to be a critical part of tender analysis and contractor appointment.

Once this exercise has been completed and the client is in possession of all of the information needed to evaluate each proposal, tenders should be ranked in a methodical and systematic way in accordance with the attributes of each bid and the client's needs.

The examination and ranking of bids will be made easier by the contract-sum analysis: a breakdown into individual sums allocated to predetermined elements or areas of the project that is provided by each tenderer of the lump-sum tender. Not only will this analysis enable the client to compare bids more accurately and identify any anomalies in tender pricing but will also be of some assistance in determining the validity of any costs submitted by the contractor for additional works or changes that may be requested during the construction phase. The required breakdown of the tender sum will need to be incorporated within the tender documentation and, eventually, the contract documentation.

A relatively simple method for evaluating proposals on the basis of ranking and weighting is given in Table 8.1. The use of such a system, which should be designed to suit the client's criteria, is essential in ensuring that the various proposals are correctly and fairly appraised and that the chosen proposal provides value for money and fully satisfies the client's requirements.

An inexperienced client may wonder why they cannot 'pick and mix' the best elements of each bid in order to formulate a more attractive proposal, and it needs to be understood that not only would this be contrary to the code of procedure, and possibly result in a breach of the contractor's intellectual property and copyright, but also could end in a joint action by the unsuccessful tenderers to recover their costs, as well as undermining the client's credibility with the local construction industry.

8.3.3 Acceptance of tenders

In Table 8.1, Bidder C has the highest weighted factor and would therefore be chosen purely on the basis of this arithmetical exercise. However, because the weighted totals are relatively close together, care must be taken not to base the decision only on these

Table 8.1 Ranking of design and build proposals

Factor	Weighting	Bid A		Bid B		Bid C	
		Rating	Result	Rating	Result	Rating	Result
Price	20	100	20.00	95	19.00	90	18.00
Design	15	90	13.50	100	15.00	85	12.75
Whole-life costs	25	85	21.25	95	23.75	100	25.00
Technical experience	3	80	2.40	90	2.70	85	2.55
Management resources	10	95	9.50	85	8.50	90	9.00
Financial resources	5	95	4.75	90	4.50	95	4.75
Safety record	3	100	3.00	75	2.25	100	3.00
Punctuality	9	85	7.65	75	6.75	95	8.55
Industrial relations	3	80	2.4	80	2.40	90	2.70
Quality record	7	95	6.65	85	5.95	100	7.00
TOTAL			91.10		90.80		93.05

factors, and possibly more subjective criteria would then be considered, e.g. performance at the interview and a 'gut feeling' of whether the client organisation can work effectively with the contracting company.

Once the successful proposal has been chosen, it will be necessary for the client and the design and build organisation to enter into one of the forms of contract that have previously been identified. In this context, it is essential to ensure that the contract documentation incorporates not only the original employer's requirements and contractor's proposals but also any amendments or additions resulting from subsequent negotiations between the two parties.

8.3.4 *Implementation of the project*

Following agreement of the contract terms and conditions and once the contract is formed, the contractor will normally require a period of time to mobilise their resources before actually commencing work on site, when they will obtain the necessary planning consents, building regulation and any other statutory approvals. They will also continue to carry out the detailed design, or enough of it to commence work on site and maintain continuity and progress, and to allow for any lengthy lead-in time for material delivery.

During this period, the client will have to ensure that detailed drawings and specifications, proposed suppliers and subcontractors and samples of, or specifications describing, materials and equipment are submitted for approval and officially authorised by the employer's agent. The employer's agent is normally a firm of consultants (architects, engineers or surveyors) who are employed to supervise the design and build contractor. The term 'employer's agent' is used in design and build projects, whereas other forms of procurement use the term 'client's representative'. This is because the contractor has much more responsibility and the supervising consultants act as the client's eyes and ears, so an agent has less power than a representative – see earlier discussions in Chapter 4, section 4.1.

The single-point responsibility for both design and construction should ensure that improved communications and teamwork will result from the use of this system, although it must be borne in mind that when using an integrated or fragmented type of contractor these characteristics may not be as evident as when a pure design and build organisation has been appointed.

Although during the construction period the contractor is theoretically responsible for monitoring progress and advising the client of any envisaged delays in completion, it would be a naïve employer who does not ensure that independent checks are regularly carried out to ensure that the contractor's progress is in accordance with the agreed programme and that the necessary corrective action is implemented should any operation fall behind schedule.

While the contractor is also responsible, under the contract, for exercising control of the quality of materials and workmanship and for the installation of quality-assurance schemes, the client can, and should, make provision for the independent monitoring of the quality of the works, although care needs to be taken to ensure that this activity is carried out in strict conformity to the terms and conditions of the contract.

It is in the employer's best interest to ensure that the arrangements made for reimbursement of the contractor during the currency of the project are such as to ensure that difficulties do not arise through disputes regarding the level and timing of any interim or other payments that are due. The most common payment procedure currently in use is the lump-sum fixed-price method, whereby the contractor undertakes to design and construct the project in accordance with the approved drawings and specification for the sum quoted in their tender, which is only adjusted when changes in legislation that could not be envisaged at tender stage are implemented and/or the client instructs variations to the brief during the design and construction period.

When using this procedure, and the JCT Form of Contract, interim or progress payments will normally be made, either on a monthly instalment basis against a measured amount of work completed or in accordance with a system of agreed stage payments entailing the agreement and reimbursement of fixed sums when specific progress stages have been reached. When this method is being used, the design and build firm applies for progress payment and the employer's agent checks the application against the amount of work completed, or the actual against planned progress, and issues an appropriate certificate to the employer confirming the amount of payment which should be made. As with other procurement routes, interim or progress payments are only considered as a 'payment on account' and any errors or omissions can be rectified in subsequent valuations and payments. There is no legal necessity to accurately value the works carried out to the nearest penny, although the client will not wish to overpay the contractor, and the contractor should be paid for work that they have properly carried out – less retention, of course.

The question of variations has been the subject of much discussion in the context of the design and build system of procurement, as the design is the responsibility of the contractor, so the client can only vary the employer's requirements, which are not at the same level of detail. However, more realistically, design and build organisations are aware of the need for flexibility to accommodate a client's particular requirements by means of variations and the desirability of providing a sufficiently detailed breakdown of their tender in the contract-sum analysis to facilitate the simple evaluation of such additions or omissions. It is essential, therefore, that clients ensure that such a breakdown is included in both tender and contract documentation and that the number of variations should be minimised. Where they are required, however, their cost and time consequences are agreed before any formal instructions are issued.

8.4 Benefits of design and build

As mentioned earlier, the three main benefits of this system are a faster total project time, single-point responsibility and expected savings in cost.

8.4.1 Time

On the question of faster total project time, a number of studies have found that, to varying degrees, design and build projects were associated with shorter overall project times than when using conventional systems, although the individual design and construction periods were often longer but can be overlapped and integrated as they are carried out by the same organisation. The reduction in time is also attributed to improved communications between the various members of the project team, the integration of the two basic functions of design and construction and the improvement in buildability and use of contractor's resources during the design periods (see Masterman 2002: chapter 5).

Projects using design and build can be up to 50 per cent more likely to be completed on time than those using conventional methods of procurement; construction can be up to 15 per cent faster; and total project times can be up to 30 per cent faster. This is the main reason why design and build has increased dramatically in usage in recent years as clients' need for early use of the facility became a primary consideration.

However, despite design and build projects generally having a better record in terms of overall completion times, their ability to satisfy clients is sometimes more variable than conventional systems. This may well be because the client has less input into the detailed design of the project or that a contractor is chosen with insufficient experience of the process, therefore producing a product of reduced quality.

Most clients consider that the single point of contact is probably the greatest strength of design and build, outweighing the likely savings in time and cost, and a high proportion of clients considered this characteristic to be their main reason for using the system. Clients should remember that single-point responsibility can only really be achieved if performance criteria are used in the formulation of the client's requirements and the contractor is allowed sufficient freedom to develop their own detailed design and is also prepared to accept the responsibility for any post-contract designs that have been produced by the client's consultants and incorporated into the final product.

8.4.2 Cost

On the question of cost, most clients are more interested in cost *certainty*, rather than the actual cost itself. For example, if a client is told the project will cost £1 million and that is well over their budget, then they won't go ahead or will ask for cost reductions in the design. However, if they are told it will cost £750,000 and it actually costs £1 million then they will clearly be unhappy, as they will not have budgeted for that sum at the outset and may not have made provision for the additional costs. Cost certainty is therefore important. An example of this point is the oncology wing at Alder Hey Children's Hospital in Liverpool, which was completed in 2005. Most of the funds for the building work were raised by the hospital itself through donations, fund-raising events and so forth, and when they had reached an amount equal to the expected cost (given by their design and build contractors), a contract was entered into and work commenced. The design and build contractors (HBG Construction Ltd, now BAM Construct UK Ltd) were left in no doubt that if construction costs exceeded the amount that had been raised, they would be given collection tins and asked to stand on the streets themselves. HBG rose to this challenge and ensured that the final account costs to the client did not exceed the original estimate.

The design and build approach enables the contractor to be more positive about the final cost to the client at an earlier stage, although cost certainty can only be fully achieved if the employer's requirements are unambiguous and comprehensive and are not subject to alteration during the construction phase.

There is some evidence to support the belief that when using this system the initial and final costs are lower than when using other methods of procurement, basically as a result of diminished design costs, the integration of the design and construction elements and the in-built buildability of the detailed design. However, care should be taken with assertions such as this as each project is individual, the designs are different, the site is different and the construction dates are often different. Therefore, over-simplified comparisons are odious, to misquote Shakespeare.

It has also been found that fragmented design builders performed badly, in terms of client satisfaction and on cost-performance criteria, compared with the other two categories of organisation. Moreover, value for money is difficult to assess because of the different methods, designs and services offered by contractors and the limited amount of information usually available at the tender-adjudication stage.

In terms of whole-life costs (WLC), or life-cycle costs (LCC) as they are sometimes known, unless the client is aware of the lifetime costs associated with the building they will be unable to properly judge the efficiency of the design using information in tenders based solely upon the construction cost. This is a major flaw in using capital cost as a major criterion of choosing a design build proposal, since lower capital costs generally have higher costs-in-use because of the lower quality and cost of materials used. As a rule of thumb, £1 of capital cost can create £5 of life-cycle costs, so the costs of the operation of the building should be an important consideration at the outset, especially in these times of environmental awareness.

8.4.3 Quality and functionality

In the early days of design and build, most of the literature did not identify levels of quality and functionality as a benefit when using this method of procurement. This result undoubtedly reflected the attitude among some clients, and certainly most architects, that design and build is suitable for simple, uncomplicated projects and the belief was that the aesthetics and quality of the finished product when using this method were lower than could be achieved by other systems of procurement.

Current experience does not support this viewpoint, as many large, complex and prestigious projects have been constructed using design and build, confirming the opinion of many that the risk of the client obtaining a crude design solution, or a substandard project, arises only if the employer's requirements are inadequate and the selection of the bidding contractors is not carried out with sufficient rigour. The system has been shown to perform better in terms of quality on complex or innovative buildings than on more simple developments. It was also established that only 50 per cent of projects using this method met the client's quality expectations, compared with 60 per cent of conventionally procured projects (see Masterman 2002: chapter 5).

The system is therefore generally considered suitable for the implementation of most types of building, provided that the employer's requirements are carefully and accurately specified. It may well be that, if the system is used with aptitude and skill by competent clients, designers and contractors, most types of new-build projects can now be successfully completed. There is now a considerable amount of advice available to both commercial and public-sector clients on how to provide a satisfactory brief and generally commission a design and build project, and there is therefore little excuse for the system not to be used successfully.

8.5 Variants of design and build

The four variants of the design build system are:

a	novated design and build	c	turnkey method
b	package deals	d	develop and construct.

8.5.1 Novated design and build

It has become increasingly common for clients to appoint consultants to carry out the conceptual design of a project and preparation of design and build tender documentation for their projects and, once the contractor has been appointed, to novate the design (i.e. ask the contractor to take responsibility for the design work carried out so far, sometimes together with the original design team) to the successful bidder to carry out the detailed design as the contractor's directly employed consultants.

The tender documentation will contain details of the client's consultants and the proposed novation procedure, together with a requirement that the contractor who is eventually awarded the contract will have to accept responsibility for the total design of the project, including the initial work carried out by others. In other words, it is as if the consultants, both legally and practically, had always been the contractor's designers. On occasions, details of the conditions of appointment of the novated consultants by the contractor, including the level of fees, have been predetermined by the client, although normally this is a matter for negotiation between the contractor and the original design consultants.

Such an arrangement should enable the design of the project to proceed more smoothly from the pre-contract to the post-contract stage, although it does mean that any pre-contract negotiations that take place between the bidders and the client will have to be carried out using the services of the contractor's 'temporary' designers, who were responsible for the outline designs contained within the contractor's proposals.

Once the client's consultants have been novated to the contractor, they will no longer be available to provide advice on the detailed design to the client, and it is therefore likely that the client will need to employ new consultants to examine the final scheme and to confirm or otherwise its suitability. Some clients may still consider that the novated consultants still represent their best interests, despite being employed by the contractor. However, this is extremely unlikely, as the designers, for both professional and practical reasons, are not likely to wish to put their relationship with the contractor at risk.

Problems can also arise as a result of the contractor being compelled to employ consultants rather than being able to choose their own designers, with whom they may well have an ongoing and successful relationship on this type of project. This kind of forced marriage may well produce a less than happy team, in which the necessary high level of integration, co-operation and collaboration between design and construction which should be inherent in the design and build system is not achieved.

8.5.2 Package deals

The package-deal system is the precursor and parent of design and build proper. As the name suggests, the intention of the original concept was that clients would be able to purchase a total package, virtually off the shelf, to speedily satisfy their building needs at the most economical price.

While the idea of being able to 'buy' a suitable building as if one were purchasing any other large consumer article is attractive in theory, in practice the fact that many package dealers merely provide an adapted standard product means that they are unlikely to satisfy fully the needs and criteria of the majority of clients.

The fundamental difference therefore between the design and build and package-deal systems is that the former method provides a bespoke design solution to suit the client's specific requirements whereas the latter uses a proprietary building system in order to produce a scheme which as stated is unlikely to satisfy all of the client's needs.

Figure 8.2 illustrates the system and the various relationships between members of the project team. Provided that the purchaser's requirements are flexible, this method can be an attractive proposition, particularly as the probable reduction in the design, approval and construction stages of the project can lead to savings in time and cost.

The majority of package-deal contractors, by their very nature, employ their own in-house designers and can thus be categorised as pure design builders; as such, they can be expected to perform well, particularly in terms of the speed and time criteria. Some of the products of this method may lack aesthetic appeal, but as the potential client is often able to see actual examples of the contractor's product before reaching a decision this potential difficulty can often be avoided. Additionally, package deals are most often used for repetitive projects of relatively simple construction, such as retail parks, and as such the architectural merit is limited.

Another aspect of the use of this particular variant is that of the proven stability and safety of the design, in that many proprietary systems have been tried and tested in use over many years and are often less prone to teething troubles than bespoke designs.

In all other respects, the package deal replicates the characteristics of the design and build system, although the forms of contract used with this method are likely to be contractor-drafted, rather than any of the nationally recognised standard forms, and

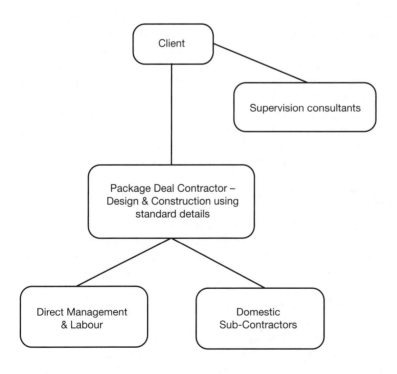

8.2 Contractual and functional relationships: package deal

clients and their advisers should therefore study the terms in detail to ensure a fair approach is taken.

8.5.3 The EPC/turnkey method

The 'engineer, procure, construct' (EPC) or 'turnkey' system is, as the name implies, a method whereby one organisation, generally a contractor, is responsible for the total project from engineering design through to the point where the keys are handed over to the client as a fully operational facility. The responsibility of the contractor is therefore often extended to include the installation and commissioning of the client's process or other equipment and sometimes the identification and purchase of the site, recruitment and training of management and operatives, the arranging of funding for the project and possibly the operation of the facility for a fixed period of time, which is twenty-five to thirty years in the case of PFI projects (see the following section and also Chapter 11).

The EPC/turnkey method was pioneered in the USA in the early 1900s, and has been extensively used there since then, by the private sector, for the construction of process plants, oil refineries, power stations and other complex production and process engineering facilities. The international standard form of contract, FIDIC, has a standard form for EPC/turnkey projects, commonly known as the 'Silver Book'. In process facilities, there is an additional variant, known as 'engineer, procure, install and commission' (EPIC), which defines each stage more closely, and although there is no internationally recognised standard form of contract, each major client, such as an oil company, will have its own standard conditions for use on their own projects.

Many UK consultants and contractors have been involved in turnkey projects overseas, particularly in developing countries, where the volume of work carried out by this method is higher than in Europe or the USA. The use of the system in the UK has been limited, and the amount of work carried out in the industrial and commercial sector appears to be small by comparison to the USA. However, the introduction of PFI by the UK government in the early 1980s resurrected a concept that has been established for many years, having been used in the nineteenth century to provide infrastructure such as railways and canals, whereby private finance was used to design, build and operate major public projects such as the Channel Tunnel, the Dartford river crossing and, more recently, various hospitals, prisons, and so on

8.5.4 PFI-type contracts

Governments throughout the world have now, for both political and economic reasons, endeavoured to cut public capital expenditure by granting the private sector a concession to finance, design, build and operate major public projects which would have in the past been carried out by the public sector using more traditional methods of procurement.

In order to attract the private sector, various methods of guaranteeing returns on investment, as well as the more mundane, but equally important, profits on the design, construction and operating elements of the project, have been devised. These arrangements are generally described as concession contracts, but are also known as BOO (build, own and operate), BOT (build, operate and transfer or build, own and transfer), BOOT (build, own, operate and transfer) and DBFO (design, build, finance

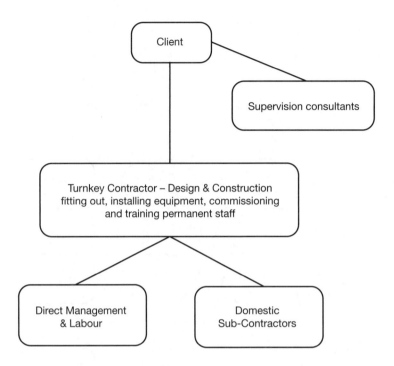

8.3 Contractual and functional relationships: EPC/turnkey

and operate). Other less common variants include BRT or BLT (build, rent/lease and transfer) and BTO (build, transfer and operate).

These various methods require that the company, or more likely consortium of companies, which has been granted the concession, design, build, operate and at some specific time transfer the facility to the commissioning client body or even retain it in perpetuity. Figure 8.4 illustrates a typical structure for a BOT project.

During the operating stage, when the facility is in use, the concession company receives income from users or the public-sector client of the facility, and in this way the public sector obtains the facility without any major capital expenditure at the beginning of the project, but instead pays for it out of revenue expenditure during its economic life.

Although many such arrangements have been implemented for major infrastructure projects, the provision of socially orientated building projects such as hospitals, prisons and so forth has only comparatively recently begun to use the PFI format. Its use was expected to increase considerably but the economic recession of 2008 and the reluctance of banks to finance major capital projects has changed the development economics considerably. Only time will tell if the method will recover.

Apart from the fundamental advantage to the client of being able to take over a fully operational facility, or, in the case of PFI schemes, reducing public-sector capital expenditure in the short term while establishing a commercially viable development in the long term, from a construction viewpoint the turnkey method echoes all of the

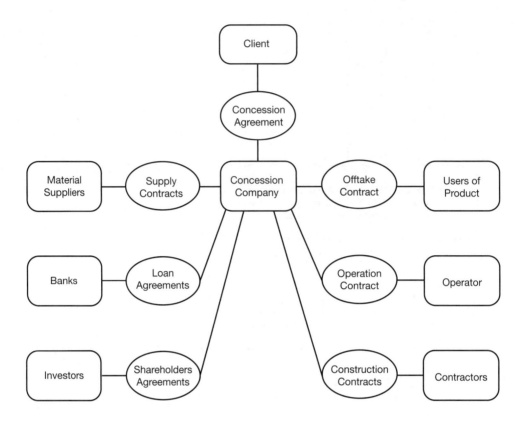

8.4 Typical structure for a BOT project

characteristics of the design and build system. The forms of contract, however, used with this variant are usually drawn from the process engineering industry rather than from the construction industry.

8.5.5 Develop and construct

Within this system, the client's consultant is provided with a detailed brief from which they prepare conceptual drawings and a site layout. The contractor develops this conceptual design, produces detailed drawings, chooses and specifies materials and submits these proposals with their bid in the same way as with design and build proper.

The use of this variant is therefore appropriate where the client wants, or needs, to determine the detailed concept of a project before inviting competitive tenders and yet still requires a single organisation eventually to take responsibility for the detailed design and execution of the project.

The main difference between design and build and this variant is the extent to which the design of the project has been developed by the client before inviting tenders. In most cases, the design will be developed at least up to outline planning stage and may, in sensitive planning locations, be taken to the point where full planning approval could be obtained.

This method is most frequently used where the client:

- employs their own in-house consultants;
- sees advantages in using a consistently retained consultant with previous experience of similar types of projects;
- may wish to limit knowledge of their intentions to a trusted few;
- wishes to minimise the differences, so often experienced at tender stage, among normal, individual design and build submissions.

It is the responsibility of the develop and construct contractor to ensure the structural sufficiency of the whole design and that the building is fit for its intended purpose. This question of design responsibility should therefore be adequately defined and covered in both the tender and contract documentation for the project.

Once again, all other aspects and characteristics of this variant echo those of the other integrated methods included within this category of systems, with the forms of contract being those previously described for the parent design and build approach.

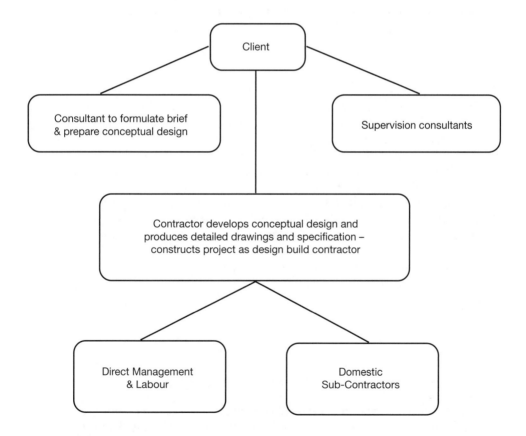

8.5 Contractual and functional relationships: develop and construct

8.6 Summary and tutorial questions

8.6.1 Summary

Common aspects

There are a number of advantages and disadvantages common to all of the methods included within the category of integrated procurement systems.

ADVANTAGES

- The single point of contact between the client and the contractor that is unique to this category of procurement systems means that the client has the advantage of dealing with one single organisation which is responsible for all aspects of the project.
- Provided that the client's requirements are accurately specified, certainty of final project cost should be achieved, and this cost is usually less than when using other types of procurement systems.
- The use of integrated procurement systems enables design and construction to be overlapped to a considerable extent and should result in improved communications being established between client and contractor. These two characteristics enable shorter overall project periods to be achieved and project-management efficiency to be improved.

DISADVANTAGES

- If, as often happens, the client's brief is ambiguous and does not communicate their precise wishes to the contractor, there can be great difficulty evaluating proposals and tender submissions.
- The absence of a bill of quantities makes the valuation of variations more difficult and restricts the freedom of clients to make changes to the design of the project during the post-contract period.
- Although well-designed and aesthetically pleasing buildings can be obtained when using this category of procurement system, the client's control over this aspect of the project is generally less than when using other methods of procurement.

Each of the individual systems within the integrated category has specific advantages, disadvantages and characteristics.

Design and build

ADVANTAGES

- Time: shorter overall project duration as a result of ability to overlap design and construction.
- Single point of responsibility – there is no passing of fault to others.
- Cost: lump sum or guaranteed maximum price allows for cost certainty.
- More controllability over costs thanks to single point of responsibility.

- Quality and functionality: better integration of specialists.
- Improved buildability of the design.
- Majority of project risks are passed to the contractor.

DISADVANTAGES

- The performance of design and build contractors is subject to considerable variation depending on whether they are pure, integrated or fragmented organisations.
- Levels of technical and managerial competence are likely to be lower as the client's choice moves from the first, through the second to the third type of contractor owing to the difference in capability among organisations specialising in design and build with in-house resources covering all disciplines (pure), a general contractor with partial in-house expertise (integrated) and a medium/small builder in consortium with an out-house design team (fragmented).
- Conversely, project costs are likely to increase as the client's choice moves from fragmented through integrated to pure design and build organisations.

Novated design and build

ADVANTAGES

- The retaining of the same design consultants through all the stages of the process should ensure that design standards are consistently maintained throughout the pre- and post-contract phases of the project.

DISADVANTAGES

- There is no guarantee that the novated consultants will be able to establish a good working relationship with the contractor and the forced marriage of the two parties may produce more problems than it solves.
- The client may also be put to additional expense in appointing new consultants to monitor the design of the project during the post-contract stage.

Package deals

ADVANTAGES

- The client is usually able to see actual examples of the package dealer's product in real situations and assess their practical and aesthetic appeal.
- Many proprietary systems have been tried and tested over a period of years and are thus likely to be free of the initial constructional defects which affect some bespoke projects.

DISADVANTAGES

- This method uses proprietary building systems to produce schemes which may not satisfy all of the client's needs.

- Some serious structural failures have occurred among some of these proprietary systems, which have also suffered from other less serious defects as a result of poor design and detailing.

Turnkey method

ADVANTAGE

- When using this system, the client is able to operate their facility and commence operation immediately upon taking possession of the project.

DISADVANTAGE

- The cost to the client of using the turnkey method can be higher than when using other more conventional procurement systems.

Develop and construct

ADVANTAGE

- This system is useful when the client has their own in-house design expertise, regularly uses external designers and sees advantages in retaining them, wishes to restrict the knowledge of their intention to build or wants to minimise the difficulties of comparing disparate design and build submissions while at the same time requiring a single organisation to take responsibility for the detailed design and construction of the project.

DISADVANTAGE

- Responsibility for the design of the project can be a possible area of dispute when using this system owing to the involvement of both the design consultants and the contractor in this aspect of the project.

The effect that these characteristics have on the selection of the most appropriate procurement system for a specific project is discussed in detail in Chapter 12, but as a general guide it can be safely stated that most new-build industrial and commercial projects – where the client's requirements can be unambiguously and comprehensively delineated, where they remain constant during the currency of the project and where they are not required to be of high architectural merit – could be procured by one or more of the integrated procurement systems that have been described.

This is not to say that aesthetically challenging and high-quality projects cannot be successfully undertaken using these systems, but rather that clients should be aware that in order to implement such projects they will need to be extremely disciplined in the way the project is managed. The establishment in 1999 of the Design and Build Foundation – a body intended to regulate design and build operators by registration, education and other means – may well help to widen the successful use of the system as well as raise standards.

8.6.2   *Tutorial questions*

1   What do you understand to be the major differences between pure design and build, integrated design and build and fragmented design and build?
2   What would you expect the following documents to include:

a   employer's requirements
b   contractor's proposals?

3   What do you understand by the term 'novated design and build'?
4   What are the major differences between a package deal and a turnkey contract?
5   Why would it be beneficial to have a design and build contract within a PFI arrangement?

9 Long-term relationships
Partnering

9.1 History and principles of construction partnering

Partnering is a relatively modern concept intended to provide a framework for the establishment of mutual objectives among the members of the building team which attempts to improve working relationships, encourage the principle of continuous improvement and develop an agreed dispute resolution procedure. All of these will benefit both the project and the participants.

This framework of mutual collaboration is designed to encourage trust, co-operation and teamwork in what is traditionally a very fragmented and adversarial process and where the combined effort of the participants should focus on the client's needs and project objectives so that the project is built as efficiently, effectively and economically as possible.

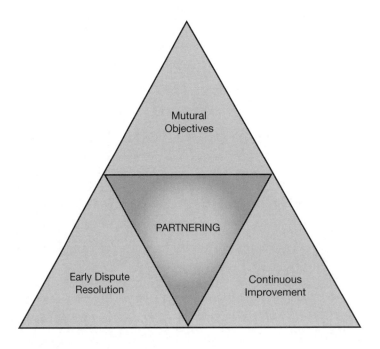

9.1 The partnering triangle

Partnering has grown out of the development of strategic alliances initiated by major construction clients in order to manage the supply chain of their projects. This concept first originated in Japan, the USA and Australia and evolved out of the perceived failure of the traditional procurement methods to meet basic client criteria and to achieve project objectives as a result of the increasing complexity and size of modern construction projects. Its foundation in Japan may well be traced to the Japanese management concept of 'kaizen' (which means small incremental improvements). However, partnering focuses more on the importance that all parties have in the construction process, and to get the critical engagement of all parties is almost a necessary condition for them to make these small improvements for the good of the whole.

Since its introduction into the UK some ten to fifteen years ago, support for the concept of partnering from major clients and government has been clear. Major client organisations with considerable construction activity, such as the British Airports Authority, Railtrack and Tesco, have all reported positive, measurable and identifiable results when using this form of procurement. However, the operational concept of partnering can be rather confusing in terms of what it is and what it is supposed to achieve.

9.1.1 Client dissatisfaction

As we have already seen, the UK construction industry has recently been analysed and observed in a number of major reports. The underlying concern of all these reports has proved to be the inefficiency of the construction industry as compared with others, together with its adversarial culture and working practices. The reasons have been given as the fragmented nature of the industry, that no two buildings are the same, the separation between design and construction, the role of consultants and the procurement methods used. While responses to all these criticisms have been made, they have not, by and large, resulted in greater productivity or customer satisfaction. Clients therefore naturally take these responses as excuses rather than reasons.

More recent reports have focused upon client needs and perceptions. Undoubtedly, both private- and public-sector clients have concluded that the construction industry needs to reflect the best practices of modern manufacturing industry to provide a satisfactory product. Indeed it was suggested in the Egan Report that the UK construction industry is seen in the same unfavourable light as the 1960s automobile industry (Department of Trade and Industry 1998).

Perhaps the Latham Report has proved to be the most significant milestone in the advent of partnering, in that it indicates for the first time that the public sector should change procedures and methods to incorporate this concept, which is used so successfully in the USA, Australia and Japan. Indeed, the USA and Australia are used as the main point of reference, having instigated suitable procedures for the selection of subcontractors in public-sector contracts (HMSO 1994a).

Partnering is still a procurement route within a commercial competitive environment, so while appropriate long-term relationships can be built up, the partners must initially be sought through a competitive tendering process, and for a specific period of time. Any partnering arrangements should include mutually agreed and measurable targets for productivity improvements in order to ensure efficiency and best value to the client.

While the Latham Report identified the major problems of the construction team and made a substantial number of recommendations, there remained a considerable number of large private-sector clients that were still to be convinced that yet another report would lead to positive change. The Egan Report represented the views of these clients and accepted, perhaps surprisingly, that the lowest price should not be the only criterion used for selection and that other factors must also be considered. This gave even more impetus to the principle of partnering.

To summarise from the Egan Report:

> [T]he Taskforce wishes to emphasise that we are not inviting UK construction to look at what it does and do it better; we are asking the industry and Government to join with major clients to do it entirely differently. What we are proposing is a radical change in the way in which we build. We wish to see, within five years, the construction industry deliver its product to its customers in the same way as the best consumer-led manufacturing and service industries. To achieve the dramatic increases in efficiency and quality that are both possible and necessary, we must rethink construction.

The Executive Summary of the Egan Report also makes the point that:

> The industry must replace competitive tendering with long term relationships based on clear measurement of performance and sustained improvements in quality and efficiency.

Much of the construction industry throughout the world is governed by the culture of competition, and in the UK this has historically been most prevalent within the public sector, which previously had been restricted by the rules of compulsory competitive tendering (CCT). The public sector (both central and local government) is still the largest client of the UK construction industry according to the latest figures. As we have seen in previous chapters, it relied mostly upon traditional methods of selecting contractors, using a traditional single-stage lump-sum tendering process and awarding the project to the lowest bidder, after a period of tender analysis.

While it would be wrong to attribute all of the failings of the UK construction industry to this process, it undoubtedly has a number of serious disadvantages for the industry and its clients. A fuller discussion of these disadvantages is given in section Chapter 6, section 6.1.

Partnering has therefore evolved, partly in response to client demand linked to the knowledge that some countries would appear to obtain better value for money than others, and there also appears to be evidence that both the public and private sectors have accepted the notion of partnering as a mechanism to achieve better results. But what then is partnering from a practical point of view and is there any evidence that it is satisfying the construction client's needs and expectations?

9.1.2 Definitions

As we have seen, the construction industry has a fragmented nature, and this is generally considered to be a management problem which could be resolved by the use of different procurement techniques. However, you cannot just change the emperor's clothes and expect a new emperor. The problems of fragmentation relate to supply of goods and services, methods of production and so forth. Looking at these piecemeal has mostly failed to address the problems of integrating the entire construction team into the construction process. Indeed, most standard forms of construction contract are adversarial by nature and still rely heavily on contractually explicit procedures rather than on mutually agreed methods to achieve financially sound objectives for all the team.

Partnering, therefore, is a device that encourages greater integration of the project team and should create competitive advantages to all that participate in the project. In the early 1980s in Japan and the USA, where TQM (total quality management) was beginning to get a hold on management thinking following research by the likes of Baldridge and Deming, team building, co-operation and equality, rather than the us-and-them relationship of adversaries to a project, were encouraged. It is perhaps from South East Asia that the concept of 'win–win', so prevalent in the cultures of China and Japan, emanates.

In the UK, partnering has steadily gained popularity from the early 1990s, and the Labour government in the first decade of the new millennium continued to support radical changes in the way that the construction industry performs and provides services to clients. One of the key features of the Egan Report is the recommendation that clients and the construction industry rely less on competitive tendering and formal construction contracts and move to a 'supply chain' system of construction production more commonly found in manufacturing industry.

To quote again from the Executive Summary of the Egan Report:

[T]he industry will need to make radical changes to the processes through which it delivers its projects. These processes should be explicit and transparent to the industry and its clients. The industry should create an integrated project process around the four key elements of product development, project implementation, partnering the supply chain and production of components. Sustained improvement should then be delivered through use of techniques for eliminating waste and increasing value for the customer.

This, it is suggested, improves productivity and therefore profits. These features are consistent with the ethos of the partnering process, which attempts to steer parties towards trust and co-operation to eliminate formal contractual and the inevitable adversarial positions and simultaneously eliminate price as the major mechanism of selection. Client organisations which have carried out projects using the partnering process all report that results are extremely positive, with large savings in terms of cost and time.

The principle of partnering has subsequently been reviewed in various reports and research projects. There appear to be more similarities than differences in opinions concerning the definition of the concept. The National Economic Development Office, in *Partnering: Contracting without Conflict* (National Economic Development Office 1991), provides a working definition of partnering as:

> *[A] long-term commitment between two or more organisations for the purpose of achieving specific business objectives by maximizing the effectiveness of each of the participants.*

In comparison, other definitions relate to the notional benefits on offer to the individuals participating in the process. These benefits generally focus on the development of long-term relationships between the parties that are based upon mutual trust and concern the implementation of mutually agreed objectives, agreed dispute resolution procedures and continuous improvement for all the participants. To this end, partnering might be considered as a concept or a notional set of abstract ideas applied to the construction process to achieve efficiency and mutual satisfaction.

Even more academically, partnering can also be considered as a 'philosophy' or soft skill of the client organisation, which has a particular system of beliefs. Partnering is therefore a concept or paradigm which has an ability to transform contractual relationships into a cohesive team with a set of common goals and established procedures for resolving disputes. This view is consistent with other definitions which are based on two essential prerequisites to partnering, mutually beneficial goals and a high level of inter-organisational trust. This must be supported by a practical mechanism that allows for both dispute resolution and performance benchmarking. But they emphasise that partnering is essentially linked to the management of people across organisational boundaries. Fortunately, modern ICT technologies assist this concept greatly, in that project extranets can be developed so that different companies working on the same project may not even know that they are different companies.

While many definitions are given for partnering, these definitions will naturally change with the evolution of the process. However, there appears to be consensus that there are some defining features of 'successful' partnering. These may be listed as follows:

- mutually agreed objectives and goals;
- inter-organisational trust;
- mechanism for problem resolution;
- continuous improvement related to benchmarking process.

In order to achieve mutual trust, organisations must ensure that individual company objectives or goals are not sought at the expense of the project objectives. An understanding of all parties' objectives is the all-important factor, but these should be aligned to the needs of the team. Of course, it is very easy to say this, but in times of economic hardship business survival is clearly the main priority and project participants will naturally seek to maximise their return and minimise their costs, as we have seen in the credit crunch recession of 2008.

Regarding dispute resolution procedures, successful partnering arrangements should provide for a pre-agreed dispute resolution process. Perhaps this is the inevitable first phase of partnering. It is difficult to eliminate the adversarial nature of the partners until a culture of mutual trust and shared gains is established.

Thus by bringing the parties to the project in a framework of trust and co-operation, the principle of partnering encourages all parties to consider continuous improvement to the work process. This has been termed 'strategic partnering', and as the growth in the use of partnering increases, this longer-term effect can be measured and noted.

Critics of the system note that of course everyone agrees that motherhood and apple pie are good things, but this is the real world and the customer is king. Customers sometimes become greedy, and it is not unusual for them to attempt to bully the contractor, subcontractors or consultants. In fact in some overseas cultures it is expected. Corporate clients often feel the need to exert their power over suppliers and consultants, and may simply pay lip service to the principles of partnering for political expediency. However, it has to be stated that most research up to the mid-2000s points away from this view, to satisfactory partnering relationships. The longer-term effect of the credit crunch recession will be interesting to see.

Many companies have indeed reported radical successes with partnering arrangements. Sainsbury's (a major supermarket chain in the UK) is an outstanding example of success in partnering, citing tremendous savings in time and cost and the narrowing down of the supply team to mutual satisfaction. Some contractors also claim that partnering is now the major basis of their turnover.

9.1.3 Cost savings and incentives

It is claimed that by using cost benchmarks, partnering can save up to 10 per cent of the original project budget, while over the longer term some suggest that partnering can result in savings of up to 30 per cent. However, it is unclear in many of these cases how the benchmarking is to be established or whether traditional methods of cost saving could have had the same result; for example, in the recent past value engineering was held to produce savings of a similar nature. Data and statistics can be used to justify almost any argument, as it often depends on how the data are presented. Nevertheless there appears to be evidence that major clients are satisfied with the financial results of partnered projects and have in some cases achieved continuous improvements, especially where the construction is of an uncomplicated and broadly repetitive nature (e.g. out-of-town retail parks, supermarket chains and fast-food outlets).

The US Army Corps of Engineers found that using partnering on both large and small contracts results in an 80–100 per cent reduction in cost overruns, virtual elimination of time overruns, 75 per cent less paperwork, significant improvements in site safety and better morale. Clearly, these figures depend on what they were comparing against.

A similar outcome was experience with the £12 million Nepean Hospital extension in western Sydney, Australia, where the project was completed early and below budget under the partnering concept. There was no time lost through industrial disputes and the site's safety record was well above the industry average.

The most successful partnering arrangements to date have looked to integrate not only consultants and main contractors, but also key suppliers. This has led to some radical changes in procurement methods involving a new approach to risk management. Some clients are even seeking to adopt forward-looking incentive arrangements, linking the success of all parties to their individual business objectives, which can extend from the existence of a simple non-binding charter which encapsulates their aims and aspirations to numerous contractual methods for aligning the strategies and objectives of partner companies.

This can be achieved by one or more of the following methods:

- *Target cost contracts*, where a target cost is agreed and the contractor is reimbursed his prime costs plus a fee. However, any overrun or underrun is shared in pre-agreed proportions (painshare/gainshare). This gives transparency and aligns the client and contractor in a common purpose.
- *Alliancing*, where the profitability of each party contributing to a project is governed by the overall success of the project and not their individual contract. It is therefore in everybody's interests to co-operate.
- *Contingency fee*, where a consultant ties a proportion of their fee to the success of the overall project. If it is not successful they will earn less than their normal fee. If it is a success, then the additional fee is financed by some of the overall savings.

Incentives can therefore be used to stimulate the project team to consider the importance of time and quality rather than concentrating on lowest cost. If this is correct, then it could be said that both clients and design teams have indeed crossed the Rubicon. As mentioned previously, the bad taste of poor quality lingers long after the sweet smell of low price has disappeared.

In some cases it may be more appropriate to use incentives to stimulate better performance in the actual use of an asset, particularly given that the average whole-life operating costs of a building are some five times the original capital costs and the facilities management contract is let using performance-based terms and conditions. With the use of techniques such as value engineering, cost planning and so forth, the capital costs of a building are as competitive as they have ever been, and Latham's 30 per cent reduction is therefore probably difficult to achieve. However, quality improvements that affect running costs are much easier to achieve by the same value-engineering principles. Site safety and impact on the environment are two other incentives that have been successfully utilised, both of which are important in modern society.

Numerous types of incentive exist for improving time, cost and performance, all of which have their advantages and disadvantages. However, the real skill lies in combining these incentives to properly reflect both the client's objectives for the project and the distribution of profit for the contractor, in such a manner that the client's objectives are achieved in a balanced way. One of the benefits of carrying out this process is that the client's objectives are analysed in detail and become very clearly defined and prioritised.

9.2 Standard forms of contract for partnering agreements

9.2.1 NEC3 Contract (Option X12)

The NEC was first issued by the Institution of Civil Engineers in 1991. Unlike the ICE conditions, it is not a joint publication by all sides of the industry. At the present time, the contract is in revision 3 – hence the title NEC3.

Objectives of the NEC

The principal objectives of the NEC are to provide a flexible set of contract documents for use on large and small projects in the UK and overseas. The complete set is contained in a series of 23 volumes and is evidence of the task which has been undertaken to achieve the objective given in section 5 of the Banwell Report, i.e. to produce a common form of contract for all construction work. It is intended for use with, or without, bills of quantities, for traditional, cost-reimbursable or management contracts. The contractor may execute no work themselves (management only) or they may subcontract any proportion up to 100 per cent. The contractor may, or may not, carry out all, or part of, the design of the works. Allocation of risks can be dealt with as required. It is therefore a very flexible procedure.

The contract is intended to promote good management, reduce the uncertainty (and therefore disputes) and minimise the necessity to consult lawyers – always a good thing!

The contract is also intended to increase clarity of requirements for each party by using words and phrases in business/conversational English rather than in complex legal drafting and by using flow charts for the procedures, which are more readily understood by construction professionals and managers.

NEC3 Option X12: partnering

By selecting this option for an individual project, all the parties to the project (realistically, all parties must be signed up including the client, consultants, contractors and subcontractors) have agreed to incorporate the partnering agreement into an existing contract as a 'bolt-on' and also signed up to the following core clause, which appears at the beginning of the document:

> *The Employer, the Contractor, the Project Manager and the Supervisor shall act as stated in this contract and in a spirit of mutual trust and co-operation.*

This, therefore, is an overriding principle of the working relationships. Many existing construction practitioners may feel a sense of cynicism, because they have seen all this before, and since the culture of the construction industry is one of master and servant all the way down the supply chain, if the client can take advantage of the contractor or the contractor can take advantage of a subcontractor then that is all well and good. This culture is exactly what these initiatives are trying to change and to move away from the view that for every winner there must be a loser.

Option X12 core clauses cover the following matters:

1   general
2   the contractor's main responsibilities
3   time
4   testing and defects
5   payment
6   compensation events

7   title
8   risks and insurance
9   termination.

Also in the document is a schedule of core-group members, which should include:

- name of partner
- name of core-group member
- address and contact details
- joining date
- leaving date
- key performance indicators
- targets
- measurement arrangement
- amount of payment if the target is improved or achieved.

Incentives. In addition to any target or bonus arrangements in an individual contract or one of the partnered company's own contracts, there is a provision in the overriding partnering contract that the partner will be paid the amount stated in the schedule of partners if a KPI is improved or achieved. In order to do this, additional KPIs may be added to the agreement.

Working arrangements. The partners are expected to work together as stated in the partnering information. They should give each other access to all information (known as an 'open-book' principle, which has become common on major infrastructure projects where partnering arrangements are used extensively). Each partner is also expected to give an early warning to the other partners of any matter that could affect the achievement of another partner. In order to make this work, it is not unusual for the partners to use common information systems.

The employer, or their representative, normally convenes and takes the chair at meetings of the core group and also prepares and maintains a timetable showing the proposed timing of the contributions of the partners. A partner gives advice, information and opinion to the core group and to other partners when asked to do so by the core group and should also notify the core group before selecting a subcontractor.

9.2.2 PPC2000 form of contract

Main features

PPC2000 is a fairly lengthy form of multi-party contract, published by the Association of Consultant Architects (ACA), for procurement of capital projects and was launched by Sir John Egan in September 2000, with a specialist form of subcontract published in 2002. It has since been adopted widely in both the public and private sectors. Its main advantage is that, like the NEC, it is written in plain English and therefore relatively easy to understand, although critics say that it is inflexible in its requirements, which does not lead to easy amendments for particular projects. Unlike the NEC, it is not a suite of contracts and therefore for something which purports to serve a cradle-to-grave requirement on projects, this one-size-fits-all concept can create difficulties.

The key differences between PPC2000 and any other published contract form are that:

- It is a multi-party concept and form of contract, which somewhat goes against the basic legal principle that a contract is an agreement between two parties.
- As a multi-party contract, there is a key concentration on supply-chain management.
- It covers the entire duration of the design, supply and construction process.
- It includes new team-based timetables, controls and problem-solving mechanisms.

Integrated team. By using a multi-party contract, this effectively puts the contractor, the consultants and any key specialist subcontractors and suppliers on the same contractual terms and conditions through a single contract, so that they are fully aware of each other's roles and responsibilities and also owe each other a direct duty of care. This is to avoid the risk of inconsistencies, gaps or duplications which can arise when contract terms are negotiated individually without overall co-ordination or ensuring back-to-back provisions and thereby establishes a much stronger legal base for all activities. It also avoids the client having to act as the conduit for communication and resolution of problems between other team members, which can occur when there is only a hierarchical structure of contractual relationships. By getting all parties signed up to a group agreement, there is going to be more direct contact between the parties themselves.

Integrated process. In order to obtain better value from projects it is now recognised that the early involvement of the contractor and any specialist contractors to the design stage is essential. PPC2000 creates the contractual structure to achieve this by providing for the contractor, consultants and any specialist subcontractors or suppliers to be appointed as early as possible in the design-development process and to work to a single integrated timetable through the design and pricing stages to commencement of the project on site. As a project management tool, PPC2000 is designed to create a clear structure and set of processes to govern the preconstruction phase of the project, which is the time when value-engineering, value-management and risk-assessment exercises will have the most impact on the design with a view to reducing or eliminating unnecessary costs.

Build-up of designs/supply chain/prices

The early creation of a team that includes all of the contractor's supply chain requires recognition of the total project budget and the contractor's level of profit and overheads. Additionally, appointing the construction team early requires the selection criteria to be qualitative (which may include some outline cost factors) rather than only a lump-sum price. In order to do this, PPC2000 provides for the following sequence of activities:

a   design development with main contractor and any specialist subcontractor input;
b   analysis of the main contractor's business cases for any single source selection (through direct labour packages or preferred specialist subcontractors/suppliers) and the open-book tendering of other subcontract packages;
c   approval of each work package and agreement of whether specialist subcontractors/

suppliers will join the overall partnering team, in either case with the approval of robust fixed prices;

d   analysis and management of risks to reduce or eliminate price contingencies;

e   incentivisation of cost savings and added-value proposals that derive from the value engineering of designs (where prices have previously been approved) or the reduction of risks (where risk contingencies have previously been approved);

f   the finalisation of an agreed maximum price supported by an open-book price framework with a complete supply chain and the satisfaction of any other agreed preconditions to commencement of the project on site.

Risk management

PPC2000 provides not only for risk analysis but also for risk management, particularly during the preconstruction phase. It provides for a review of each relevant risk with proposals for its elimination, reduction, insurance, sharing or apportionment as appropriate. It also provides for the notification of any proposed pricing of risk contingencies with proposals for their removal or reduction.

Project on site

As PPC2000 commences during the design stage and covers the total project, the same procedures during the construction phase. Of particular importance during this stage are:

- an early-warning system regarding any problems in performance;
- advance evaluation of any proposed change or any event of delay or disruption and a restriction on the contractor's right to obtain additional profit or head-office overheads as a result of delay or disruption (effectively the earlier involvement of the contractor in an ordered process through to start on site is a trade-off for excluding their right to profit from later claims should there be any problems on site);
- operation of a core group of key individuals representing partnering team members, who are the medium for adding value through a partnered collaborative approach – if they can reach agreement (if they cannot, the project proceeds on the basis of the agreed documentation);
- a contractually binding project timetable governing the interface between team members during the construction phase, which follows on from the partnering timetable that governs those activities during the preconstruction phase;
- agreed incentives including financial links between achievement or non-achievement of agreed KPI targets;
- a structured approach to alternative dispute resolution including a problem-solving hierarchy and reference to the core group and to conciliation or mediation;
- the use as appropriate of a partnering adviser to support the entire partnering team (rather than an individual member of it) in documenting their relationships and advising on the new relationships and processes in practice.

Comparisons to other procurement methods

A number of the innovations in PPC2000 were already present to some degree in the NEC suite of contracts (e.g. early-warning requirements and advance evaluation of compensation events). However, PPC2000 takes these much further and seeks to achieve a level of integration not present in NEC or any other form of contract. It tackles directly, through full integration and scheduling of activities, the following risks that exist in two-party contracts:

- protracted design development being solely in the hands of consultants without any contractor involvement;
- inadequate information issued to tendering contractors at tender stage, which often results in tenders being loaded as an allowance for risk;
- inadequate time for tendering, which can have the same effect as above;
- hidden information regarding the relationship between the contractor and specialist subcontractors/suppliers (by way of discounts, etc.). This hidden profit is more difficult to come by in open-book relationships;
- unwillingness of project participants to declare problems early and propose solutions, because of an 'it's-not-my-problem' attitude and possibly inviting claims;
- absence of advance information in relation to changes or delay/disruption, to enable the client and other project participants to mitigate their effect;
- absence of binding schedules, with resulting misunderstandings and consequent delays. This of course has a flip side in that binding schedules as a contract document can lead to the client being restricted in issuing variations because of possible extension of time claims;
- absence of alternative ways of resolving disputes, thus encouraging the risk of adjudication, arbitration or litigation.

PPC2000 in practice

PPC2000 was adopted initially by housing associations and local authorities (primarily on housing schemes). It has since spread to other types of project and other sectors, including local and central government programmes for schools, hospitals, leisure facilities and other public buildings. It has also started to be used for highways and engineering works and has extended to private-sector clients in the newly privatised industries, such as railways.

Projects within the UK (see Trowers and Hamlins 2005) that have adopted PPC2000 include:

a   the Whitefriars Housing Group's £240 million programme in Coventry, which has achieved significant reductions in time and cost savings compared to previous procurement methods;
b   the £8 million Raleigh Square project for Metropolitan Housing Trust to provide environmentally friendly offices and housing in Nottingham;
c   a £40 million programme of integrated refurbishment and repair works for Welwyn Hatfield Council, which has achieved cost and time savings recognised in the district auditor's reports;
d   a £1.34 million British Aerospace (BAe) project for office and production facilities;

e   a £5 million tower block refurbishment by Places For People in Newcastle, where the partnering contract enabled them to overcome at minimum cost/delay the insolvency of a specialist cladding subcontractor;

f   the £5 million Watergate special needs school in Lewisham in north-east London, which has been accepted by DfES as a benchmark project;

g   a £575 million programme of public buildings works and highways and engineering works set up by Durham County Council;

h   station refurbishment and retail projects undertaken by Virgin Trains;

i   a programme of new-build housing undertaken by Western Challenge Housing Association, where the use of PPC2000 enabled the team to survive the insolvency of a constructor.

PPC2000 is therefore designed to achieve greater integration and better results in the procurement of any project. It requires and rewards closer client involvement in an integrated project process leading to:

- removal of gaps or duplications between project-team members and avoidance of confusion which often leads to wasted time and money resolving these at a later stage;
- clear timetables through to start on site with the resulting savings in time and cost;
- earlier contractor involvement leading to better buildability and efficiencies with the potential to improve quality/reduce cost;
- more open cost information to establish price accuracy, reduction or removal of arbitrary price contingencies and closer control over the consequences of variations, changes and unforeseen events;
- improved performance of all parties through early creation of a partnering team supported by improved communication and mutually compatible roles and responsibilities.

9.2.3 Public Sector Partnering Contract (PSPC)

The PSPC suite of partnering contracts is intended to provide a legal framework that fully complements the partnered approach to construction project delivery.

According to the PSPC website: 'The contract's flexibility, simplicity and ease of use enable project teams to integrate and partner effectively, while the "user-friendly" content allows team members to operate efficiently without the extensive legal constraints that are imposed upon them under standard forms of construction contract.'

The production of the PSPC suite of contracts was the responsibility of the Federation of Property Societies (FPS). The FPS represents all property-related service groups in UK local government and was formed in the 1980s at about the time that the Latham and Egan Reports were being published. The FPS recognised the need for a legal framework that fully supported the partnered approach to project delivery. It envisaged a suite of contracts that embodied both a sound legal platform and a flexible structure. This would allow project teams to build robust partnering procedures and processes specific to their projects. To achieve this, the FPS sought the expertise of James R Knowles Ltd. Roger Knowles, then Chairman of James R Knowles Ltd and a well-respected figure in legal services within the construction industry, set to work. He developed the Public Sector Partnering Contract through consultation and in collaboration with senior partnering and procurement specialists from other associated firms.

The resulting suite of contracts was designed to provide the structured legal framework necessary to protect the interests of project-team members whilst promoting a partnering approach compliant with the Egan agenda. The contract is designed to enable a collaborative approach to project management which involves the agreement of processes and procedures to meet the needs and the objectives of the project in hand. It seeks to avoid pre-defining any procedures that may not be suitable for every project or party, which is often the case with other standard forms of contract. It is this blend of legal content and procedural empowerment that makes the PSPC contract suitable for local government projects.

The PSPC contracts also have a number of core benefits that differentiate them from other standard forms of contract:

- *Flexibility*. The different options available account for all project types, sizes and circumstances.
- *Usability*. The contract is concise and straightforward, and is accessible to all project-team members.
- *Simple terminology*. The PSPC seeks to avoid the use of legal terminology, so that all members of a project team can easily read and understand it. This increased understanding is designed to support the ethos of teamwork, reduce misunderstandings and limit the likelihood of potential disputes.
- *Cost transparency*. The target cost option allows for full open-book accounting processes in order to promote effective value engineering by the project team.
- *Improved working practices*. Using the target cost option means that team members can be incentivised to improve performance through shared savings.
- *Best-practice driver*. By promoting the partnered approach to project delivery, more efficient and cost-effective working practices are developed by project teams.

Partnering agreement

All contracts require execution of a partnering agreement, which overarches the individual construction contracts. The partnering agreement is quite short, comprising only eleven terms and conditions, and contains a partnering charter, which requires all parties to engage in collaborative team working in a spirit of 'mutual trust, good faith and co-operation', as stated in the partnering agreement.

The agreement also requires common information systems to be set up on the projects and for all parties to work in the best interests of the project. As with other contracts, such as the NEC and PPC2000, an early-notice requirement for potential problems is important for each party. Ambushes and a spirit of 'it's not my responsibility' will not be tolerated from any of the parties.

Procurement options within the PSPC suite

As mentioned previously, the PSPC is a suite of contracts, suitable for several different procurement routes and contract strategies. There are eight different options under the contract, depending on whether the client decides to take responsibility for the design themselves or delegate it to the contractor and whether payment to the contractor is based on a lump sum negotiated at the pricing stage, or a cost-reimbursable method. Table 9.1 gives a clear picture of the main contract options 1–6, together with the two subcontract options.

The main options have 35 to 38 clauses, depending on the option, and set out duties clearly in plain English. The suite seeks to keep the main benefits while avoiding the traps of other contracts. Notably, there is no duty to trust or co-operate and there is no partnering adviser.

On the plus side, the contract administrator role is unambiguous and the contract is deliberately written with back-to-back risks, roles, responsibilities.

PSPC Professional Services Contract

There is also a standard form of consultant's contract with PSPC for use by independent architects, engineers, project managers and quantity surveyors. This is almost a shell contract, as the fee structure and payment terms are totally flexible and there is no consultant delivery programme. However, the professional services contract does require all information which is held by the authority to be provided to the consultants, which is now a major requirement of health and safety legislation.

Pre-start agreement

When the contractor is appointed before the pricing stage of the project, i.e. the contractor's costs, overheads and profit have not yet been calculated, it is not unreasonable to pay them for their input in the same way that professional consultants are paid for their design stage input. The pre-start agreement is therefore used for early contractor involvement and includes items such as programme requirements; payment based on costs only, i.e. no profit; the provision of records; a definition of the services required by each party and any outline cost estimates already produced.

9.3 Summary and tutorial questions

9.3.1 Summary

It would appear that many within the industry have accepted the challenge and have entered into partnering arrangements willingly, although with some natural reservations. These arrangements are still be in the stage of evolution and can only be achieved through experience, although how the credit crunch and the recession of 2008–10 will affect attitudes remains to be seen.

Table 9.1 Procurement options under the PSPC suite of contracts

	Term maintenance (planned and responsive)	Client responsibility for design	Contractor responsibility for design	Subcontracts
Lump sum	Option 1	Option 3	Option 4	Option 7
Target cost/cost reimbursable	Option 2	Option 5	Option 6	Option 8

In summary, partnering arrangements are based in whole or in part on the following:

- *Value-based procurement*: value can only be truly assessed by the client, therefore the client's principal criteria must be set out in the tender documents, together with their relative emphasis or ranking, to enable transparency and engender trust.
- *Single-point responsibility*: the fragmentation of design and construction is the major cause of project ambiguities, which can lead to disputes. It is therefore recommended to use a design build form of contract.
- *Inter-organisational partnership*: a partnership between organisations to achieve common goals is essential to success.
- *Means of dispute or issue resolution*: a non-adversarial dispute resolution procedure should be adopted and agreed by all parties.
- *Continuous improvement*: all partnerships should have a requirement for continuous improvement.
- *Longer-term relationships* instead of proceeding project-by-project. Dealing with a customer over the medium to long term has greater benefits as a result of shared experience and knowledge (see Chapter 10).
- *Mutual gains* for all participants to the process, i.e. win–win culture.

The jury is still out on the various claims about cost savings and benchmarking, but partnering as a procurement process is here to stay. The concept of total project costs and whole-life costing are now quite mature in their processes and should enable the project to be designed and built as effectively as possible, given the modern requirement for low operating costs, carbon offsetting, etc.

The benefits and advantages of a partnering approach have been neatly summarised in Table 9.2 (from Matthews, Tyler and Thorpe 1996).

Table 9.2 Benefits and advantages of partnering approach

Trait	Client	Main contractor	Professional consultants	Partnering subcontractors
Improved team approach	✓	✓	✓	✓
Improved understanding of project		✓		✓
More compliant bids		✓		✓
High-quality bids		✓		✓
Better/closer relationships	✓	✓	✓	✓
Better/more reliable programming	✓	✓		✓
Better way of achieving project objectives	✓	✓	✓	✓
Limited competition		✓		✓
Less confrontation (claims)	✓	✓	✓	✓
Lower tendering costs	✓	✓		✓

9.3.2 *Tutorial questions*

1   The Egan Report, which was published in 1998, wished to see the construction industry change to a predominantly partnering approach within five years. Has this been achieved? If not, what do you think are the main reasons?
2   Any initiative which reduces costs and increases efficiency must be a good thing. Does partnering do this?
3   The three forms of partnering contract are now well established. Compare the main features of each.
4   Is partnering suitable for all construction projects, irrespective of client type and size? Give reasons for the answer.
5   What do you understand by an 'early-warning system' in this context?
6   What do you understand by 'open book'?

10 Long-term relationships
Framework agreements

10.1 What is a framework agreement?

A framework agreement is simply a time-limited (term) contract between a client and a supplier (contractor) that governs and overarches individual project agreements awarded under it during the framework term. It may be used where a client body, knowing it has a number of construction projects to implement over a certain period, decides to carry out a single tender process for both the consultants and contractors that it will need for the projects. Alternatively, it may be used by a client who wishes to develop a stronger relationship with a particular single contractor or consultant across a number of its projects. While the framework agreement is generally between the employer and the supplier (contractor), there is a positive encouragement for the supplier to enter similar framework agreements with its own suppliers and subcontractors, thus including the entire supply chain within the overall framework.

For the client, the framework agreement has the obvious primary benefit of reducing tender costs. If it can carry out one tender exercise for ten projects, rather than tendering each project separately, there will naturally be both a cost and a time saving.

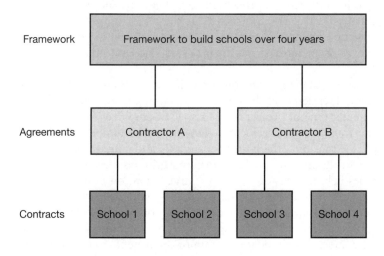

10.1 Example of framework agreements

Source: www.constructingexcellence.org.uk.

There is also the possibility of economies of scale. The framework agreement will often require the supplier (contractor or consultant) to include in the overall tender a price mechanism for the individual project contracts that may be made under the framework agreement (termed 'call-off'). Where the contractor or consultant is likely to obtain a steady stream of work over a period of years, this learning curve is expected to be reflected in the price. The framework agreement will normally require the contractor to improve their performance over time as they become more familiar with working with the client or on a particular type of project. This performance will be monitored and judged by key performance indicators (KPIs), or alternatively, a percentage reduction in preliminary costs over a number of projects may be required or negotiated. As in any relationship, communications are expected to improve over time as each party becomes more familiar with how the other works and what is important to them. While this can be obtained through individual contracts over a period, the improvement should be more structured and measurable within a more formal framework agreement.

For the consultant or contractor, it is undoubtedly beneficial to have a guaranteed work stream and income over a period of time. This allows better forecasting and planning for resources and staffing as well as providing a buffer against difficult times. Where framework agreements are put together on Egan principles, the relationship should be much less adversarial with fewer disputes, allowing all parties to concentrate on the project and, hopefully, lead to fewer calls to claims consultants, lawyers and the professional indemnity insurance provider.

An initial criticism of framework agreements was that they result in contractors and consultants becoming complacent in that they have a cosy relationship with the client and therefore quality and professional standards may slip, which of course starts from the spurious assumption that quality and standards are guaranteed to be better in competitive tendering. Most clients are aware of this danger, any complacency would quickly manifest itself with the KPIs and framework agreements normally contain a get-out clause for such eventualities. Another criticism is that the client may be denied the benefit of a fresh, innovative approach. In fact, clients are more likely to get innovative approaches from framework contractors as the majority of research and innovation in the construction industry is carried out by contractors, so not spending money on tendering leaves them more funds to innovate, increase efficiencies and hence drive down costs.

A further criticism is that framework agreements will reduce competition. Unless you are one of the successful tenderers at the outset, your company may be precluded from working on a particular type of project or for a specific employer for the framework period. Where the framework agreement covers a greater number of projects over a period, it is arguable that only the larger consultants and contractors are able to pre-qualify for such a substantial amount of work and fully resource the projects, leaving smaller companies behind. This depends to a large extent on how the framework agreement is set up. Some employers prefer to contract with a small number of contractors or consultants for a larger number of projects, but not all. Others may appoint, say, ten companies to their framework agreement, but only intend to procure six projects under it. Of course, this eliminates certain of the benefits to the employer in terms of improved efficiency. It is also increasingly difficult to persuade contractors to bid for such projects as they become disillusioned with being appointed to a framework but not necessarily winning any work.

Some contractors in particular have complained that employers have unrealistic expectations of the savings that a framework agreement can bring. An intention to reduce time and costs over a number of projects is laudable but imposing this on the contractor or consultant as an obligation is, perhaps, unfair.

10.1.1 Is a framework agreement a contract?

The Public-Sector Procurement Directive of the European Union (EU) covers public procurement of services, supplies and works. If a framework agreement is not in itself a contract for services, but merely an overarching agreement covering such contracts, then the Directive will not apply. The UK has always taken the view that the only sensible approach to such framework agreements is to treat them as if they are contracts in their own right for the purposes of the application of the EU rules. Therefore, in the UK, the practice has been to advertise the framework itself in the *Official Journal of the European Union* (*OJEU*) and follow the EU rules for selection and award of the framework. This provides transparency and removes the need to advertise and apply the award procedures to each call-off under the agreement, on the basis that the framework establishes the fundamental terms on which subsequent contracts will be awarded.

The Directive defines a framework agreement as

> *an agreement with suppliers, the purpose of which is to establish the terms governing contracts to be awarded during a given period, in particular with regard to price and quantity.*

In other words, a framework agreement is a general term for agreements with providers which set out terms and conditions under which specific purchases (the call-offs) can be made throughout the term of the agreement.

The framework agreement may itself therefore be a contract to which the EU procurement rules apply. This would be the case where the agreement places an obligation, in writing, to purchase goods, works or services in exchange for payment. For this type of agreement, there is no particular problem under the EU rules, as it can be treated in the same way as any other contract.

However, the term is normally used to cover agreements which are not, themselves, covered by the definition of a contract to which the EU rules apply (though they may create certain contractually binding obligations). Such agreements set out the terms and conditions for subsequent call-offs but place no obligations on the procurers to buy anything.

Therefore, in EU Directive terms, with this approach contracts are formed only when goods, works and services are called off under the agreement. The benefit of this kind of agreement is that, because authorities are not tied to the agreements, they are free to use the frameworks when they provide value for money, but to go elsewhere if they do not.

It is this form of agreement, where the framework itself cannot be readily classifiable as a contract for the purposes of the current Directives, which has caused much difficulty in relation to the application of the EU procurement rules, and which is addressed explicitly in the new provision. But it should be stressed that the contractual status of a framework agreement should not cause undue concern. The key is that a means of awarding contracts under framework agreements is provided for without the need to re-advertise and re-apply the selection and award criteria from the outset.

10.2 Setting up a framework agreement

It will be important for each individual client to consider whether a framework agreement is the right approach for the particular set of projects under consideration. This will generally be a value-for-money judgement for the client body, taking account of the kinds of projects involved and the ability to specify their requirements with sufficient precision at the outset. In particular the framework should be capable of establishing a pricing mechanism which will be valid throughout the term of the agreement (possibly five to ten years). However, this does not mean that actual prices or rates should always be fixed, but rather that there should be a mechanism that will be applied to pricing the works during the period of the framework. It should also be possible to establish the scope and specification of the works that will need to be called off. There should not be any objection to upgrading the works required so long as it remains within the scope of the original specification.

In using framework agreements, clients will need to ensure that their obligations on issues such as sustainability, TUPE and Code of Practice on Workforce Matters such as health and safety are being met. The use of framework agreements does not remove the need to address these issues, where relevant, in awarding a contract at the call-off stage.

If the framework approach is chosen, as mentioned above it will be necessary to advertise the framework itself in the *OJEU*, if its estimated maximum value over its lifetime exceeds the relevant EU threshold and the procurements in question are not covered by one of the exclusions set out in the Directives. If the framework itself is not advertised in *OJEU*, in cases where the projects are subject to the EU rules, an *OJEU* notice may be required for individual projects and call-offs. The need to do this will depend on the size of the call-offs and on whether the aggregation rules apply. It is far better, therefore, to advertise the framework itself, so that there is no need to go through the *OJEU* procedure as each individual project call-off comes up.

Clients which act as central buying organisations may set up and advertise framework agreements on behalf of other clients, whether in their own overall organisation or subsidiaries. Where the EU rules have been followed by such central buying organisations, other clients may use the framework agreements as required so long as they have been covered in the *OJEU* notice. The new provision explicitly recognises that contracting authorities may purchase through central purchasing bodies.

It is probably clear by now that framework agreements are really only suitable for very large clients with a rolling programme of projects and where the client's requirements can be given as a performance specification, leaving the detailed design and construction to the contractor. Most framework agreements in the UK are in infrastructure projects and utilities, such as water/sewage treatment plants, pipelines, etc. In these cases, the high-level performance specification (e.g. capable of treating x cubic metres of water per day) is relatively straightforward.

Under the new Directive the *OJEU* notice must make it clear that a framework agreement is being awarded and include a list of the authorities/clients entitled to call off under the terms of the framework agreement. The authorities can be individually named, or a generic description may be used: water authorities, health authorities, etc. When generic descriptions are used, it is advisable to include, in the notice, a reference to where details of the authorities covered can be obtained. Although the individual circumstances will need to be considered, it is worth seeking to construct the framework so that it can have the maximum take-up across the public sector. However, if the framework is relevant only to, say, certain central government departments, that should be made clear.

The *OJEU* notice should also state the length of the framework agreement. Under the January 2006 Directive, it will be a maximum of four years 'except in exceptional circumstances, in particular, circumstances relating to the subject of the framework agreement'. A longer duration could be justified in order to ensure effective competition in the award of the framework agreement because four years would not be long enough to provide a return on investment to the contractor, especially for specialist engineering works which may take several years to complete. It is worth considering, in any event, whether a framework agreement is necessarily the best vehicle for a longer-term project.

The notice must also include the estimated total value of the goods, works or services for which subsidiary projects or call-offs are to be placed and, so far as is possible, the value and frequency of the call-offs to be awarded under the agreement. This is necessary in order for contractors to be able to gauge the likely values involved and to provide a figure for the framework overall which, as with other contracts, should not normally be exceeded without a new competition taking place.

Once the *OJEU* notice has been despatched, the authorities setting up the framework agreements should follow the rules for all phases of the procurement process covered by the Directives. This will include the use of the open or restricted procedures or, where the conditions for their use are met, the negotiated or competitive dialogue procedures, and adherence to the rules on specifications, selection of candidates and award. The Directive does not explicitly prohibit the possibility of concluding framework agreements under the new competitive dialogue procedure. However, it is difficult to imagine cases where the conditions governing the use of a competitive dialogue would be satisfied and the use of a framework agreement would be practicable. At the award stage, the providers to be included in the framework agreements should be chosen by applying the award criteria, to establish the most economically advantageous tender or tenders, in the normal way.

Framework agreements can be concluded with a single provider or with several providers, for the same programme of projects. In the latter case, there must be at least three providers, provided that there are sufficient candidates satisfying the selection criteria and which have submitted compliant bids meeting the award criteria. The agreement will establish the terms which will apply under the framework, including delivery timescales and rates.

10.3 Call-offs or subsidiary/underlying contracts

When a client enters a framework agreement, it can appoint the contractors or consultants on the framework for individual projects, without further recourse to the EU procurement procedures. This should be either by means of simply engaging one of the framework contractors or consultants without further competition, providing this procedure is documented in the framework agreement, or holding a mini-competition between those on the framework capable of performing the contract.

Where several clients wish to create a joint framework, they can carry out a single tender exercise as long as each client is named as a contracting authority in the relevant documentation.

The duration of subsidiary projects or call-offs under a framework agreement is not specifically limited by the Directives. For example, call-offs for consultancy services might be for three, six or twelve months or longer and for major construction projects could be five to ten years, as mentioned above. As a result, it may be the case that individual call-offs extend beyond the term of the framework itself, although this should not be done in order to circumvent the EU rules. The length of call-offs should be appropriate to the purchases in question and should reflect value-for-money considerations.

Where a framework agreement is concluded with just one provider, call-offs under the agreement should be awarded on the basis of the terms laid down in the agreement, refined or supplemented by other terms in the framework agreement but not agreed at that time. It is the same principle as that applying to a normal contract, except that, with a framework agreement, there will be an interval between the awarding of the framework itself and the calling off of the individual projects under it. There can be no substantive change to the specification or the terms and conditions agreed at the time that the framework is awarded.

Where frameworks for the same programme of projects are awarded to several providers, there are two possible options for awarding the call-offs under the framework.

10.3.1 Call-offs by applying the terms of the framework agreement

Where the terms laid down in the framework agreements are sufficiently precise to cover the particular call-off, the authority can award the call-off without reopening competition. The Directive does not specify how this should be done, although in order to ensure value for money, the authority should award the call-off to the provider who is considered to provide the most economically advantageous (value for money) offer based on the award criteria used at the time the framework was established. In the event that the contractor is not capable or is not interested in the project, the authority should turn to the next-best provider. For example, frameworks might be concluded with five providers for the building of new primary schools in an education authority's district within a set period. When the client education authority wishes to build the school in y town or x village, it would go to the contractor offering the most economically advantageous offer, using the original award criteria for that project alone, without reopening the competition. If that provider for any reason could not resource the project at that time, the authority would go to the provider offering the next most economically advantageous offer, and so on.

10.3.2 Call-offs by holding a mini-competition

Where the terms laid down in the framework agreements are not precise enough or complete for the particular call-off, a further or mini-competition should be held with all those contractors within the frameworks capable of meeting the particular need.

This does not mean that basic terms can be renegotiated, or that the specification used in setting up the framework can be substantively changed. Substantive modifications to the terms set out in the framework agreement itself are not permitted. It is more a matter of supplementing or refining the basic terms to reflect particular circumstances for the individual call-off. Examples of such terms are:

- particular delivery timescales;
- particular invoicing arrangements and payment profiles;
- additional security needs;
- incidental charges;
- particular associated services, e.g. installation, maintenance and training;
- particular mixes of quality systems and rates;
- particular mixes of rates and quality;
- where the terms include a price mechanism;
- individual special terms (e.g. specific to the particular products/services that will be provided to meet a particular requirement under the framework).

Where a mini-competition is held for a particular project call-off, the client authority should consult in writing (i.e. invite to tender) the providers within the framework which are capable of meeting the particular need. This does not necessarily mean that every provider in the framework must be included, as a framework may cover a number of different supplies or services, and there is no obligation to consult those providers which had not originally agreed to provide the particular project which is the subject of the call-off. Indeed, the framework agreement itself may be divided into categories, each covering different contracts or services. In that case, the authority need only consult providers in the categories which cover the projects required.

However, there is no scope, at this stage, to run a selection procedure based on technical ability, financial standing, etc. This is a pre-qualification matter and should have been carried out before the framework itself had been awarded and should not be repeated at the further competition stage. The decision on who should be consulted should be made on the basis of the kinds of projects or services required and which providers can supply them, based on their offers at the time the framework agreement itself was awarded.

Client authorities should state the subject matter for the call-off for which tenders are being requested, and also a time limit which is sufficient to enable the selected contractors to submit their bids for the particular call-off. This time limit should take account of the complexity of the call-off and the time needed for the different tenderers to submit their bids. In addition, where the authority has decided to make use of the option to hold an electronic auction for the mini-competition, it must abide by the rules covering e-auctions as set out in the Directive. Tenders should be submitted in writing, and they should remain confidential until the time limit has expired. The client authority should award the individual project to the provider which has submitted the most economically advantageous tender on the basis of the award criteria set out in the framework itself, focusing on the particular requirement. New award criteria should not be added, but weighting may need to vary to reflect the particular requirement.

10.4 Standard forms of framework agreement

There are four major standard forms of framework agreement in use in the UK:

1   NEC3 Framework Contract (FC)
2   NEC3 Term Services Contract (TSC)
3   JCT05 Framework Agreement (FA)
4   NHS ProCure21 Framework Agreement.

10.4.1 NEC3 Framework Contract

The framework contract within the NEC family of documents is designed to allow the employer to invite tenders from suppliers to carry out work on an 'as instructed' basis over a set term. Normally, the employer will appoint a number of suppliers within the framework agreement and instruct them individually to carry out work within the defined scope of the original agreement.

When the employer decides to develop a framework arrangement, they will need to define the scope and the end date of the agreement term. The scope identifies the extent of work that is covered by the contract and may refer to a specific location, the type of work to be carried out or other relevant information which would normally be included in a client's brief. The scope should be sufficiently defined that both the employer and supplier(s) are satisfied that each of the suppliers has the resources and capability to carry out the works likely to be instructed.

Work under a package order can be carried out under the conditions of any of the primary and secondary options of the NEC. The NEC Professional Services Contract could be used for design and advisory services, and the NEC Engineering and Construction Contract or the NEC Engineering and Construction Short Contract for works.

The employer will select the most appropriate form for work under the framework, depending on the type and complexity of the work to be carried out. The employer prepares documents for inviting tenders for the overarching framework contract and these are set out in the contract data for the framework contract and will include the following:

- *The framework information*, which contains information about the management of the framework contract. It will include the need for co-ordination meetings and other commitments which are not paid under a time charge order or a package order.
- *The contract data applicable to all time charge work* – the appropriate entries from the contract data for the NEC Professional Services Contract conditions.
- *The contract data applicable to all work under a package order* – the appropriate entries from contract data for the selected contract conditions.
- *The selection procedure* – how the supplier for a work package is to be selected.
- *The quotation procedure* – how the supplier and employer are to prepare and assess a quotation for a proposed work package.

The employer must also describe the information required from tenderers for preparing a quotation – the quotation information – and other material needed to assess tenders or manage the framework contract. Tenders are submitted, assessed by the employer and a number of suppliers are selected and appointed under the framework contract.

When the employer has identified the need for some work within the scope of the framework contract (i.e. a call-off project) the supplier for the work will be selected using the 'selection procedure'. If a contractor's design advice is needed before the work can be fully defined, the selected supplier can be instructed to help using a time charge order.

When a time charge order is issued, the employer must include the further contract data needed to define the work to be carried out – further material in the 'scope' under the Professional Services Contract. In addition, other elements from the NEC Professional Services Contract data, such as starting date and completion date, must be given where relevant. The work under a time charge order is carried out under the NEC Professional Services Contract conditions, using the options identified in the contract data. The effect of the time charge order is that the work instructed is incorporated into the framework contract.

When the work is sufficiently defined, the employer will seek a quotation from the supplier, which should be prepared using the quotation information data provided at tender and this quotation must comply with the quotation procedure set out by the employer in the original framework contract. This will set out the basis for preparing the quotation and explain the amount of detail needed by the employer to assess the quotation.

When they receive the quotation, the employer can take a number of actions, depending on their objectives. If they have received an acceptable quotation, it can be accepted by issuing a package order to the supplier. If the quotation is not acceptable, the employer may change the work package and ask for a revised quotation. The employer is not required to accept any of the quotations and may decide that the work does not need to be carried out at all.

The work under a package order is carried out under the NEC conditions of contract identified in the contract data. Again, the effect of the package order is that the work instructed is incorporated into the framework contract.

Further work can be instructed until the end date, unless termination takes place at an earlier date. Either party is entitled to terminate at any time, which means that no further work can be ordered under the framework contract, but any work already ordered as a call-off under either a time charge order or a package order will continue and can only be terminated under the contract conditions governing that work.

After the end date, unless otherwise agreed, no further work can be ordered. Work already instructed, under either a time charge order or a package order, continues until it has been completed.

10.4.2 NEC3 Term Services Contract (TSC)

This contract should be used for the appointment of a supplier to maintain a service or manage and provide a service; therefore it is not really a contract for the procurement of building work as such. These services may have discrete packages of project works or include elements of design and relate to physical works or soft services such as facilities management. The TSC contains a call-off facility, therefore can be considered as a framework agreement.

For smaller contracts, the Term Services Short Contract is an alternative to TSC and is for use with contracts which do not require sophisticated management techniques and comprise relatively straightforward services with low risks for both client and contractor.

The TSC is essentially different from other forms of contract in the NEC family. It has been designed for use in a wide variety of situations and is not restricted to construction. It is essentially a contract for a contractor to provide a service rather than a project (and is not limited to a professional or construction service) to an employer over a specific period of time. The service may be provided continuously over the period of the contract or on a task-by-task call-off basis.

10.4.3 JCT05 Framework Agreement (FA)

As with the contracts mentioned above the JCT FA is not intended as a stand-alone contract but rather as an overarching agreement for use with project-specific contracts – called 'underlying contracts'.

Clause 3 of the JCT FA Guide states:

> JCT FA has been designed for use by anyone who anticipates procuring a significant volume of work and/or services over a period of time and who wants to see a collaborative approach to such work and progressive and sustainable improvements in the way such work and services are performed.

Clause 5.1 of the Framework Agreement itself gives the objectives of the framework as:

- zero health and safety incidents
- team working and consideration for others
- greater predictability of out-turn costs and programme
- improvements in quality, productivity and VFM
- improvements in environmental performance
- right first time with zero defects
- avoidance of disputes
- employer satisfaction
- enhancement of service provider's reputation

which are clearly very good objectives to strive for.

The JCT Framework Agreement had two forms in the 2005 edition and can either be binding on both parties or non-binding, although the 2007 edition was published only in the binding form and the non-binding form was not republished. The 2007 form encourages communication of clear organisation structures and accurate briefing with those responsible and also includes provisions dealing with collaborative values which require the parties to express to personnel their desire to work in a collaborative manner.

Some weaknesses of the JCT are that the new risk-management proposals are unclear and the Framework Agreement does not offer an agreed financial incentive.

Furthermore, the JCT may have caused problems by placing collaborative processes in the FA and overriding provisions in the underlying contracts. Some contractors have expressed the view that the FA still requires significant development and integration with other contracts.

10.4.4 NHS ProCure21

ProCure21 is a framework procurement process that may be adopted by UK National Health Service (NHS) clients to deliver a wide variety of construction-related services. ProCure21 provides these clients with the ability to appoint pre-qualified principal supply-chain partners (PSCPs) within a pre-agreed commercial arrangement negotiated within the initial framework contract. This is designed to enable the individual project team to immediately focus on the needs of the NHS client.

The PSCPs are very different from traditional contracting organisations as their supply chains contain a broad range of expertise from construction professionals through to specialist medical organisations. This should provide the NHS client with the opportunity of engaging the PSCP to undertake a wide variety of duties from service strategies, estates strategies, business planning, developing the brief and design development through to major and minor construction works.

ProCure21 is based upon a long-term framework agreement of five years with provisions for extension between the Department of Health and a number of framework partners. NHS clients may select any one of the PSCPs based on past performance and track record. As the framework partners are already appointed to the *OJEU* framework, the selection procedures do not need to be repeated, thus saving the NHS client both time and money.

ProCure21 is therefore designed to offer flexibility in terms of the level of service provided depending on when the PSCP becomes involved. Their input can vary from assisting the NHS client with its overall strategy to that of developing a detailed design, depending on how early the PSCP is appointed.

An additional role is that of the implementation adviser, who provides a source of guidance to the NHS client throughout the process. As a member of the ProCure21 team, they will assist the NHS client in establishing the most appropriate application of P21 for the particular client's needs. The implementation adviser is provided by the central government's Department of Health along with a central team to administer the programme and provide performance management of the PSCPs and general advice and guidance.

ProCure21 embraces the principles of collaborative working and is designed to ensure that teams work together effectively. It also encourages the early involvement of the PSCP and supply chain to ensure that the design, cost and programme are all achievable and represent best value for the NHS client.

ProCure21 is therefore not just designed as a construction solution. A key benefit is the diverse range of management and construction services it can offer. The PSCPs act as 'solution providers', and their services may range from service strategies, estate strategies and business planning through to design development, development of the brief and construction works (both major and minor).

10.5 Summary and tutorial questions

10.5.1 Summary

The advantages and disadvantages of framework agreements include:

Advantages

- Single procurement process covering a term of years, therefore client advertising costs lower.
- Long-term relationships – designed to build trust and good working practices.
- Improved efficiency over time, therefore less waste.
- Mature performance measurement systems – will be developed over time.
- More potential for continuous improvement.
- Lessons transferred from project to project.
- Efficient teams kept together.
- Continuity of work for supply side.

Disadvantages

- Difficult for new firms to get work as all projects for several years will be given to a small number of companies within the framework agreement.
- Difficult for smaller firms to get work as series of projects make resourcing only viable for large companies
- High set-up costs for both client and contractors.
- Usually open-book accounting – client knows all the contractor's costs and margins.
- Slower to get new projects on site.
- Unrealistic expectations by some clients.

10.5.2 Tutorial questions

1   Read the article on the BAA framework agreement taken from *Building* magazine at the height of the credit crunch recession in May 2009. Comment on the reasons for this decision by BAA.
2   Read the case study on framework agreements: Building Schools for the Future.

   a   Has the arrangement provided value for money to the public authorities? Is there any evidence of this?
   b   Has the speed of delivery of the schools been better than if they were procured separately in competition? Is there any evidence of this?
   c   Has the quality of the schools improved, in terms of defects, etc.?

By Joey Gardiner

Airports operator to use competitive tenders for all schemes over £25m as it tries to bring down costs

Airports operator BAA is to stop using its £6.6bn construction framework on major projects as part of an overhaul of its procurement strategy.

The firm, responsible for the construction of Terminal 5 and developments around Heathrow, Gatwick and Stansted airports, will now put all projects worth more than £25m out to competitive tender.

The move follows a review by Steve Morgan, the firm's capital director since January. Sources close to the process said the decision was motivated by a desire to cut £200m from the firm's £6.6bn five-year construction plan.

Under the shake-up, only jobs of less than £10m in value will be allocated under the framework. For jobs worth £10–25m, the framework members will have to go through a mini-bidding process. Any jobs over £25m, such as the forthcoming £100m contract for Terminal 2b, will be procured through open tender.

According to a source close to the situation, the change could also affect some projects already in procurement that do not have a tight timescale for completion.

'We've got as much chance of benefiting as losing out' (Framework contractor)

Morgan is also believed to be seeking to make procurement less informal by ensuring no work is begun or prices agreed until full contractual terms and conditions are agreed and signed.

The five-year framework, signed last year, includes Balfour Beatty, Carillion, Costain, Ferrovial Agromán, Laing O'Rourke, Mace, Morgan Ashurst, Skanska and Taylor Woodrow.

Source: *Building*, 22 May 2009.

Case study of framework agreement: Building Schools for the Future

Extracts from the Executive Summary of National Audit Office Report dated February 2009

In 2003, the (UK Government) Department for Children, Schools and Families (the Department) announced the Building Schools for the Future Programme (BSF), which aims to renew all 3,500 English secondary schools over the 15-year period 2005–2020, subject to future public spending decisions. It plans to entirely rebuild half the school estate, structurally remodel 35 per cent, and refurbish the rest. Refurbishment includes providing new ICT to recently built schools. Local Authorities are responsible for commissioning and maintaining the schools. The Department created Partnerships for Schools (PfS) to manage the programme centrally.

1   *The Department sees BSF as important to improving educational attainment and the life chances available to children, by providing educational, recreational and social environments that support modern teaching and learning methods. It wants buildings to be shared and used by local communities, and to be flexible in responding to developing needs. It also wants BSF to support local reorganisation of secondary schools to reflect demographic needs and a greater diversity of provision, including Academies and specialist schools.*

    i   *targeting funding to groups of schools to allow Local Authorities to plan strategically for the provision of school places and other facilities, and for the delivery of children's services, on an area wide basis;*

    ii   *long-term partnering efficiencies between the public and private sectors, usually through the establishment of local joint ventures called Local Education Partnerships (LEPs), which have exclusive rights for 10 years to deliver new and refurbished school facilities and related services; and*

    iii *central programme management, coordination and support for local strategic decision making and school building and refurbishment projects.*

2   *The Department provided £3.6 billion of capital funding up to March 2008 (£2.3 billion under signed PFI contracts and £1.3 billion under conventional funding). It has allocated another £7.5 billion up to March 2011, and plans to provide further funding after that. BSF accounted for 22 per cent of England's expenditure on school buildings in 2007–08. BSF has not been included in the Government's acceleration of education capital funding to act as a fiscal stimulus.*

3   *Approximately 75 per cent of Local Authorities that had signed contracts before December 2008 have developed BSF projects under PFI arrangements. Over the course of 2008, difficulties in the banking sector reduced the amount of money available for banks to lend and it became difficult for Local Authorities to find lenders of senior debt for PFI deals. Kent County Council agreed a BSF PFI deal in October 2008 and between then and the start of February 2009 everyone that signed BSF contracts used conventional funding. The Department, PUK and PfS believe at present that BSF remains one of the more attractive markets for bidders, but the extent to which financing difficulties will have an impact on the programme as a whole is as yet unclear. The Treasury, Department and PfS are seeking new sources of private finance, including the European Investment Bank.*

Key findings
Progress in the delivery of the programme

1   *The Department and PfS were overly optimistic in their assumptions of how quickly the first schools could be delivered, leading to unrealistic expectations. In February 2004, the Department said that it wanted to build 200 schools by December 2008, but Local Authorities only managed*

to build, remodel and refurbish 42 schools through BSF. The Department underestimated the time needed to establish the programme, carry out strategic planning and procure private sector partners to build the schools. It took over a year for the Department, PfS and Local Authorities to establish the details of the programme, including the scope, overall level of funding available and the funding mix. It took Local Authorities nearly six months longer on average than initial estimates to procure a LEP, although this was a little less time than it took them on average to procure a contractor in previous school PFI projects. It also took Local Authorities about 18 months on average to develop strategic plans, compared to initial expectations of just over six months. After seeing the first few plans, the Department asked Local Authorities to spend more time to improve their proposals, because it believed it was more important to improve the quality than to accelerate the programme. PfS has streamlined the strategic planning and procurement processes so that it should be quicker in future.

2   The programme now includes the majority of Local Authorities, but scaling it up to deliver all 3,500 new or refurbished schools will be challenging. As at December 2008, PfS is working with the majority of Local Authorities to develop their schools. Fifty-four schools are due to open in 2009 and 121 in the following year. To start all secondary schools by 2020, the number of schools in procurement and construction at any one time will need to double over the next three years. Consequently, there will need to be an increase in the availability of procurement and project management skills, which are in short supply at present.

3   There has been an increase in estimated total costs. The Department and PfS estimate the total capital cost of the programme will be between £52 and £55 billion, a 16 to 23 per cent real increase from previous estimates. The majority of the increase is because the Department has increased the scope of the programme and has agreed to provide additional funding for the inclusion of Academies, Special Education Needs facilities, Voluntary Aided schools and carbon reduction measures. About a third of the increase in the estimate is because the original estimate assumed building costs would rise with general inflation, but building cost inflation is now estimated to have been twice general inflation up to 2008. To meet these costs and accelerate the programme to start all schools by 2020, annual expenditure on the programme would need to increase from £2.5 billion a year to between £3.4 and £3.7 billion a year at current prices from 2010/11 onwards.

4   The total capital cost of each BSF school averages £1,850 per square metre, which is similar to most other schools. It is less than Academies built before their integration into BSF, which averaged £2,240 per square metre at 2007 prices. The prices of BSF buildings have been kept under control by the funding arrangements put in place by the Department and implemented by PfS. These place the cost of increasing the scope of school projects with the Local Authorities and require them to keep projects affordable.

Local delivery arrangements

5   BSF is making it easier for Local Authorities to use capital funding strategically. More than 75 per cent of Local Authorities in our survey said it was leading to more strategic procurement. All of the seven Local Authorities in our case studies have put in place plans to re-organise their school estate in a coordinated way, and devoted significant time and resources to planning the investment. Initially, planning processes and guidance did not focus on the practical matters that would help schools meet expectations. The Department and PfS have improved the processes and guidance significantly for more recent projects.

6   The costs of establishing the first LEPs have been high. We estimate that for the first fifteen LEPs, the combined total cost of designing the first few schools, procuring a private sector partner, and setting up the LEP averaged between £9 million and £10 million. A large proportion of this cost was for the design of the first schools. These total costs were higher than they needed to be because of avoidable delay, extensive reliance on consultants by Local Authorities, large numbers of sample schemes and alterations made to standardised documents. PfS has started to streamline the process of establishing a LEP to reduce costs in future.

7   It is too early for Local Authorities to be able to tell if the expected benefits of the LEP model will be realised. A quarter of Local Authorities in our survey anticipate that there will be benefits from a LEP approach. But most have not yet reached the stage of developing new projects following the establishment of the LEP and consider it too early to tell. The private sector partners surveyed by the National Audit Office are more optimistic: nearly 70 per cent believe that the LEP model can offer value for money.

8   Early evidence shows that having a LEP can lead to time and cost savings on repeat procurements, although most Local Authorities have not reached this stage. The first few projects developed after LEPs were established have been procured more quickly and efficiently than comparable projects undertaken without using a LEP. In the case of Lancashire, for example, two PFI schools were procured in 12 months and 7 months, compared to the 20 months it took to procure the LEP, and half the time that was previously typical for school PFI procurement before BSF. The first non-school project delivered through a LEP was in Leeds, and was procured six months more quickly than Leeds had previously managed without its LEP. The main factors were quicker scoping and agreement of projects, which also resulted in approximately a 20 per cent saving (£200,000) on the Local Authority's internal procurement costs compared to similar procurement without a LEP.

9   The first LEPs found it difficult to establish effective working arrangements and relationships between Local Authorities and private sector partners. Governance and contractual arrangements are complex, requiring early attention to how to manage the operational phase. PfS, Local Authorities and bidders initially paid insufficient attention during the procurement process to how LEPs would work in practice. Tensions from the negotiation process sometimes adversely affected relationships when the project

moved from procurement to operation. Confusion around the scoping process and shortcomings in partnering have led to some avoidable delays and reduced efficiency in the LEPs' development and scoping of their first projects. In 2008, PfS started to focus on helping LEPs overcome these issues. Local Authorities and private sector partners are working to overcome early problems and some are starting to see the benefits of effective partnering, such as more effective town planning applications through the pooling of expertise.

10 LEPs develop projects without competitive tendering during a ten-year exclusivity period. The exclusivity arrangements could make it harder to price projects economically, as the private sector partner will not typically need to demonstrate efficiencies by competing against rivals. To mitigate this risk, Local Authorities will therefore seek alternative sources of assurance over the value for money of individual project budgets proposed. The forms of assurance can include comparison to national benchmarks and to the original cost schedules put forward by contractors for the projects developed when they initially competed to join the LEP. There is also provision for market testing after five years. In addition the contracts include continuous improvement targets, which require reduced prices for future projects, and loss of exclusivity rights for failure to deliver value for money. Public sector membership of the LEP Board also improves the transparency of costing.

National coordination of BSF

11 The Department's decision to establish PfS has helped to achieve effective programme management. PfS provides national leadership and is able to carry out programme management activities which the Department and Local Authorities could not carry out by themselves. PfS provides skilled specialist people that the Department would find difficult to recruit. It has also exercised effective control over the overall scope, flow and cost of the programme in a way that could not be done by individual Local Authorities. PfS provides structured programme management and practical support to Local Authorities, including standardised documentation and guidance, and facilitates learning from experience between Local Authorities. Its overall costs, combined with those of the Department, are comparable to other programmes with central administration of devolved capital spending, such as the Department of Health's Local Improvement Finance Trust programme, the Housing Corporation's Affordable Housing Programme and the Department for Environment, Food and Rural Affairs' Waste Infrastructure Development Programme.

12 PfS's corporate targets emphasise the timeliness of delivery. These influence performance bonuses received by PfS staff of up to 20 per cent. Although PfS's guidance and review of Local Authority plans highlight the importance of the quality of the schools being built, 70 per cent of the corporate targets are weighted towards timeliness of delivery. The Department and PUK are developing an additional set of quality performance indicators to use in future.

13 *The benchmarking tool developed by PfS to help control capital costs needs to be developed further so it is useful to all Local Authorities. PfS has developed a benchmarking tool for cost and price data to help Local Authorities gain assurance on value for money, given the ten-year exclusivity period of the LEP. It has been used where competition has been weak, but cannot as yet provide a benchmark for every Local Authority because it holds insufficient data. Effective use of the benchmarking information by Local Authorities will be essential to ensuring prices remain economic in the absence of competition.*

14 *PUK's role in helping to fund and manage PfS has resulted in higher rewards for PUK than it would get from a straightforward fee arrangement, although it also results in greater commitment and in-depth support to the programme. The funding arrangement is complex and exposes PUK to some of the programme's risks, particularly delay. PUK's return on its contribution is up to 13 per cent a year, assuming there are no delays or performance deductions.*

Conclusion on value for money

15 *This report focuses on the efficiency and economy of procurement under BSF as it is too early to measure BSF's effectiveness in improving the quality of education. The main challenges to securing value for money revolve around increasing the pace of delivery; securing adequate cost assurance; and managing relationships in a complex delivery chain, requiring buy in from a wide range of public and private sector parties.*

16 *Original expectations of how quickly schools could be built were overly optimistic. PfS will find it very challenging to include all 3,500 schools in BSF by 2020. To do so, it would need almost to double the number of projects in BSF over the next three years.*

17 *The cost of the programme has increased by 16 to 23 per cent in real terms to between £52 and £55 billion, in large part because of decisions to increase its scope but also because of increased building cost inflation. The Department and PfS have taken measures to help control capital costs so that BSF school capital costs are similar to most other school building programmes and cheaper than Academies built before their integration into BSF.*

18 *Achieving value for money through a LEP requires cost savings over the expected ten-year flow of projects to offset high initial costs. Procuring a LEP takes a long time and is costly. Costs have been higher than they need be (£9 million to £10 million to procure a private sector partner and design the first projects) and can be reduced for LEPs procured in future. There is some early evidence that LEPs can lead to time and cost savings once they have been set up, but very few Local Authorities have reached this stage. Contractors' ten-year exclusivity for developing projects within the LEP is a potential challenge in maintaining effective cost control and realising cost savings, requiring effective use of benchmarking, continuous improvement targets and market testing to gain assurance on the value for money of each project.*

19 *National coordination by PfS has brought benefits to the programme. At the local level, there is evidence that the benefits of strategic funding and central programme management are being achieved in many cases. Achieving the potential long-term partnering benefits through the complex LEP model requires clear responsibilities, accountability, commitment and buy-in from all parties.*

Recommendations
The pace of delivery and cost assurance
The Department and Partnerships UK agree PfS's corporate targets annually, which influence the size of the bonus pool available to senior staff at PfS.

So far these have focused on the timeliness of delivery, which, although important, needs to be balanced with maintaining the affordability of the programme and achieving effective outcomes. The Department should establish a smaller balanced scorecard of performance indicators for PfS than it currently uses. These should better reflect the objectives of BSF, covering the timeliness, cost and quality of the programme's outcomes.

PfS's benchmarking data will be essential to help sustain value for money for schools not procured in competition.

PfS should speed up its collection of cost information on BSF schools including procurement, capital, facilities management, ICT, life cycle costs and PFI contract variation costs, and make this information available to Local Authorities so they can benchmark their costs.

The Department should invite Local Authorities to provide detailed cost information on major school projects procured outside the BSF programme so that PfS can include this cost information within its benchmarking.

The costs of setting up a LEP have been high for the first Local Authorities to do so. These costs should fall for future projects.

PfS should monitor the costs of establishing and using a LEP; disseminate good practice; streamline and standardise the process so as to help Local Authorities to cut these costs; and use frameworks where sensible to make procurement quicker.

The complex delivery chain
A general lack of skills in procurement and programme management across the public sector constrains capacity in BSF. PfS currently helps improve skills levels on an ad hoc basis. Skilled resources, which are in short supply in the public sector, are required if the complex BSF model is to deliver the desired benefits. PfS should establish a strategy to increase the skills available to BSF. This strategy could include (i) the provision of training (potentially through contractors); (ii) shifting the balance of its own recruitment by taking on more junior staff and training them with a view to movement into Local Authorities; and (iii) facilitating the secondment and placement of skilled individuals between Local Authorities.

Many Local Authorities remain to be convinced of the benefit of the LEP approach. Poor planning for how to manage contracts during their procurement and difficulty in establishing effective working arrangements and relationships have slowed the speed at which the first LEPs are delivering their next phase of schools. The Department and PfS should obtain buy in from Local Authorities for the agreed procurement approach. The Department should encourage PfS and BSF to work jointly to promote the effective operation of LEPs and help Local Authorities manage the transition from the procurement to the operation of the contracts and the ongoing contractual arrangements. PfS should satisfy itself that all deals have arrangements in place for the effective management of contracts before approval of their Final Business Case.

Monitoring of whether local and national objectives are being achieved is unsystematic, and plans for achieving them lack detail.

The Department and PfS should:

A   *require Local Authorities to introduce a consistent system to record and monitor the full list of benefits desired for each BSF school and project, keep that system up to date and use it to track and help realise these benefits.*

B   *provide support to Local Authorities and schools in realising these benefits through, for example, developing the existing guidance on change management plans to include monitoring of who is responsible for achieving each benefit, how it will be measured, how it will be achieved, progress towards achieving it, and when it is achieved.*

C   *review the achievement of the benefits in one and three year post occupancy reviews.*

11 Public-sector projects

This chapter will discuss the changing nature of public-sector procurement in relation to the building sector. Public-sector projects are introduced with reference to traditional and current procurement processes. Contemporary integrated public–private procurement processes in the public sector are then outlined with reference to the Gateway process as developed by the independent office of HM Treasury, known as the Office of Government Commerce. The advent of the Gateway process is drawn from developments within the Private Finance Initiative (PFI). The initiative was introduced in 1992 to radically overhaul arguably inefficient and wasteful wholly public-sector procurement. The PFI method of financing the building sector and its procurement practices in public projects is discussed in this chapter in relation to procurement use within Public–Private Partnerships (PPPs). PPPs form the basis of public-sector businesses that embrace the advantages and disadvantages of both sectors. The future of PPPs are also discussed in this chapter, especially if it can be argued that this type of partnership will become a less or more frequent option for public-sector building projects and procurement.

11.1 Introduction to public-sector projects

Public-sector projects can be simply defined as all projects that are under ownership by central and local government. Public-sector projects are often criticised for having excessive 'red tape' or bureaucratic regulations that slow down completion and increase costs. Effective procurement in the public sector is therefore critical to ensuring that projects go ahead. Those projects considered effective will be those offering value for money to the public purse and that continue to meet the public interests for which projects are built. Most companies bidding for public-sector work will encounter the formal procurement process. However, it is the more complex projects that create real challenges, as they often involve an assessment that combines cost, value and risk.

EU Directives are also important to public-sector procurement interests, as the EU drives towards an open and competitive pan-European marketplace. European Directive 2004/18/EC came into effect in 2006 with objectives that sought to: promote an efficient, non-discriminatory selection process; treat applicants on a fair and equal basis; achieve value for money; and achieve transparency and accountability for bidders and for the awarding of authority's stakeholders. Effective public-sector procurement for nation states within the EU is important as it can make huge savings that can be used to reduce the public-sector borrowing requirement (PSBR) or be used to fund more public-sector projects. Public-sector construction spending excluding professional fees and VAT totals about £33 billion, 31 per cent of the total spend on all construction in the UK. Furthermore, a significant proportion of private-sector expenditure related to PFI and PPP is also derived from public-sector procurement.

The assessment process in procuring public-sector projects is vital to ensure that effective procurement is achieved. Most projects have a pre-qualification questionnaire (PQQ) that is submitted by those wishing to submit for tender. This is used to arrive at a shortlist of tenderers that are all equally capable of delivering the project. The assessment should be undertaken using only the following criteria: financial standing, technical ability, resources and legal compliance. To illustrate this, Table 11.1 outlines a typical hierarchy of assessment in pre-qualification decision-making.

Following the pre-qualification of potential tenderers, the second stage of the tender appraisal commences with the issue of an invitation to tender (ITT). Simple competitions will be based on lowest price assuming other criteria such as technical compliance, resolution of bid qualifications and financial appraisal are met. More commonplace are integrated solution competitions that assess not only on lowest price but in combination with other proposed attributes such as design, construction, operation and service provision. Integrated solution headings can include whole-life cost, financial and legal compliance, design, management, innovation and bidder-specific risk. Successful tenders should ideally go through a rigorous evaluation system. Table 11.2 outlines nine key success factors to consider in making the final selection. These are not in any order of importance and include success factors such as defining deliverables in a project (factor 1) and identifying elements of a project that could quickly stop a project from running (factor 9).

Developments in appraisal and evaluation like these have influenced elements in a public-sector procurement process that has become more intricate and highly professionalised. It has to meet both national and European Directives, and has recommended guideline structures for successful tendering. This level of sophistication in procurement is extended to the formation of government departments such as the Office of Government Commerce (OGC) that have been created in response to meeting greater purchasing efficiency and value for money. The formation of the OGC Gateway process has been one such technocratic response.

Table 11.1 Hierarchy of pre-qualification assessment

Stage 1	Evidence of technical capability to deliver the full extent of the contract. This is assessed on a simple yes/no basis.
Stage 2	An assessment of economic and financial standing. Technical review of accounts, finance ratios and value of the project relative to others in the tenderer's portfolio. Evidence of prohibited activity such as convictions for fraud.
Stage 3	Technical capability. A quantitative and qualitative assessment of staff and technical resources, the proposed team, relevant experience and management systems.
Stage 4	Final shortlisting. The final assessment involves the ranking of qualifying candidates on the basis of scores against value-added project criteria such as the readiness of the team, the understanding of requirements and the team's record of problem-solving and innovation.

Source: *Building* 2009.

Table 11.2 Nine success factors for a successful evaluation system

Factor number	Success factor
1	Well-defined deliverables, represented in bids in a structured format so that a confirmation of completeness and a like-for-like assessment can be made.
2	A fully documented appraisal system setting out the methodology used, allocating responsibility so that a consistent approach is adopted by the appraisal team.
3	Evaluation criteria that relate specifically to business needs, such as quality, management, delivery and innovation.
4	Active management of stakeholder technical assessors to ensure that assessments are carried out effectively.
5	Calibration of weightings so that one criterion is not over-emphasised at the expense of others.
6	Clear qualitative scoring systems that reduce opportunities for subjective assessment by focusing on compliance with client requirements.
7	Scoring systems that ensure the preferred solution meets the basic functional requirements and the budget, as well as performing well against other criteria. Innovation and over-compliance should be rewarded only if related to business need.
8	Scoring systems that provide some indication of value for money by comparing quality and lifetime benefits with whole-life costs.
9	A mechanism for dealing with 'showstoppers' at any stage during the assessment so that problems can either be resolved quickly or a tender can be withdrawn.

Source: *Building* 2009.

11.2 OGC Gateway process

The Office of Government Commerce (OGC) is an independent office of HM Treasury, established to help government deliver best value from its spending. OGC is therefore heavily involved in the procurement of goods and services in the building sector. Improving effectiveness and efficiency are important elements of procurement and are promoted through several mechanisms. One particular mechanism is via Procurement Capability Reviews that assess current capacity and capability in central government departments. The Procurement Capability Reviews also assess the provision of contract databases and frameworks that generate products and services designed to improve procurement efficiency. Furthermore, the reviews foster a skilled and professional network in the government procurement service while providing clear policy guidance and best practice. One specific model for reviewing best practice endorsed by the OGC is the Gateway process.

The Gateway process as promoted by the OGC examines programmes and projects at key decision points in their life cycle (OGC 2010). The significance of the Gateway process in building procurement is that it is a mandatory process in civil central government. As best practice, the Gateway process is used in public-sector areas such as health, local government and defence. The types of projects and programmes that are in the Gateway process are presented in Table 11.3. These wide-ranging types of projects and programmes demonstrate the universal appeal of the model. Furthermore, the separating out of property/construction developments demonstrates that building procurement is significant enough to warrant standardised professional practice processing methods.

Table 11.3 Projects and programmes reviewed in the OGC Gateway process

1   Policy development and implementation
2   Organisational change and other change initiatives
3   Acquisition programmes and projects
4   Property/construction developments
5   IT-enabled business change
6   Procurements using or establishing framework arrangements

Source: Office of Government Commerce 2010.

The Gateway process looks ahead to provide assurance that the programmes and projects can progress successfully to the next stage. As previously mentioned, a major element of the Gateway process is the use of reviews. For instance, independent peer reviews are made at key decision points during the life of a project. These reviews can then assess whether a project is utilising all of the best-practice guidance available and how it should move forward to be more successful. Methods for these Gateway reviews typically involve a series of interviews and documentation reviews. The review is often enhanced with the team's experience to provide valuable additional perspective on the issues facing the project team. It also provides an external challenge to the robustness of plans and processes. To provide more structure, these review methods sit against a rigid Gateway framework of review types from 0 to 5 (Table 11.4) that ensure clear and distinct process reviews have been covered.

The Gateway process is also used within local government partnerships. For instance, local government project delivery specialists use the process while working in partnership with all local authorities. This could be in securing funding and accelerating the development, procurement and implementation of certain business arrangements. Examples of these arrangements include private finance initiative (PFI) schemes, public–private partnerships, complex projects and programmes. PFI is now discussed as a specific financing option that is used in the building procurement process.

11.3 The Private Finance Initiative (PFI)

PFI is defined as a modern form of procurement to encourage private investment in public-sector projects. It was introduced as a government programme under the Conservatives (1979–97) that required all new public-sector capital expenditure proposals to be tested to see if private-sector suppliers could replace them. This programme was set against the backdrop of continued government objectives of decentralisation and liberalisation of markets. It also held the belief that private businesses are innately better than the public sector at decision-making, cost-reduction, risk-taking and organisation (Mumford 1998). Decentralisation has also been part of a wider process that goes beyond policy objectives and which can be seen in the spread of downsizing and outsourcing. In addition, developments in information technology have contributed to its advance. When the New Labour administration took office in 1997, support for PFI projects continued, with acceptance that there should be private-sector participation in public projects. This support for PFI is particularly emphasised in education infrastructure and the Building Schools for the Future

Table 11.4 OGC Gateway process review stages

Review stage number	Review type	Review output
0	Strategic assessment	This is a programme-only review that investigates the direction and planned outcomes of the programme, together with the progress of its constituent projects. It is repeated over the life of the programme at key decision points.
1	Business justification	This first project review comes after the Strategic Business Case has been prepared. It focuses on the project's business justification prior to the key decision on approval for development proposal.
2	Delivery strategy	This review investigates the Outline Business Case and the delivery strategy before any formal approaches are made to prospective suppliers or delivery partners. The review may be repeated in long or complex procurement situations.
3	Investment decision	This review investigates the Full Business Case and the governance arrangements for the investment decision. The review takes place before a work order is placed with a supplier and funding and resources committed.
4	Readiness for service	This review focuses on the readiness of the organisation to go live with the necessary business changes, and the arrangements for management of the operational services.
5	Operations review and benefits realisation	This review confirms that the desired benefits of the project are being achieved, and the business changes are operating smoothly. The review is repeated at regular intervals during the lifetime of the new service/facility.

Source: Office of Government Commerce 2010.

programme. It continued to be a mainstay of public procurement for New Labour with 750 deals with a combined capital cost of £5 billion being signed by March 2007 (*Building* 2009).

The purpose of PFI is to persuade the private sector to put capital into public services at a time when the PSBR reached £50 billion in 1992. PFI as a form of partnering has a high profile, representing public and private sectors collaborating over some of the biggest projects now being constructed. Building projects in PFI tend to be large-scale initiatives in the public domain such as highways, hospitals, prisons, schools, airport extensions, power schemes, railway lines, water-treatment plants and government offices. Since its launch in November 1992, over 400 PFI contracts have come into force representing future commitments of around £100 billion (NAO 2001: 1). In effect, PFI procurement creates a far better level of communication between public-sector clients and private-sector contractors, as both sides are linked by a common objective. The collaborative agreement provides mutual benefits to both parties. It should be noted that joint working between sectors has a reasonably long history and is not entirely a new phenomenon.

Under PFI the purchaser does not immediately acquire the assets but puts in place a contract to buy services from the supplier. This means that a supplier, or group of suppliers, acquires the assets involved (e.g. builds a school) while being secure in the knowledge there will be a customer for its services (e.g. local education authority). The assets concerned may remain with the supplier for a long period of time that may last up to thirty or forty years. Depending on the contract, the assets may then be transferred to new owners such as a public-sector agency or to a new private operator. Typically, PFI suppliers are contracted not only to finance and build a facility but also to design and operate it. The supplier may additionally provide infrastructure and support services for a period of several years following construction. The forming of a PFI can therefore be a complex process, and Table 11.5 illustrates the steps that can be taken to provide guidance through a PFI project.

Arguments in favour of PFI are that it is a way of keeping down taxation by encouraging more efficient private enterprise. Alternatively, it can be argued that PFI is simply a quick trick to cut spending and remove public expenditure from the PSBR. Further arguments against PFI raise the issue of the absence of insider knowledge for suppliers aiming to meet the service needs of those procuring the product. For instance, it can be questioned whether a supplier of schools has sufficient insider knowledge and experience of what the local education authority needs to design, build, operate and finance a successful school. The premise for using PFI is therefore complex, but its merits have not caused differing political groups to disregard it as a procurement tool. The overall argument for using PFI is still at present a minor issue, with PFI constituting only a tiny fraction of public-sector capital expenditure at approximately 3 per cent of construction expenditure in 2009 (ONS 2009).

Table 11.5 Step-by-step guide to the PFI process

Step sequence	Task
1	Clarify objectives by establishing business need
2	Appraise the options
3	Produce outline business case and outline PSC (Public Sector Comparator)
4	Assemble project team
5	Decide tactics for selection stage
6	Invite bids by issuing contract notice in OJEC (Official Journal of the European Communities)
7	Prequalification of bidders
8	Selection of bidders (shortlisting)
9	Reappraise business case and refine the PSC
10	Invitation to negotiate
11	Receipt and evaluation of bids
12	Selection of preferred bidder and final evaluation
13	Award contract and financial close
14	Manage contract

Source: HM Treasury Taskforce 2000.

11.3.1 Construction procurement under PFI

A construction project procured under PFI is based on a new kind of relationship between a public-sector client and a private-sector contractor. As briefly mentioned, the general procedure for building procurement under PFI involves contractors, usually operating in a consortium, agreeing to *design, build, finance* and *manage* a facility traditionally provided by the public sector. In return, the public-sector client agrees to pay annual charges during the life of the contract and/or allows the private sector to reap any profits that can be made for a specified period, which may last up to thirty years or more. In this way, both sectors can be seen to be specialising in what they do best: with the public-sector client setting the agenda by specifying the level of service required and the private-sector contractor determining the best way to deliver that service. These PFI arrangements have obvious advantages, as the contractor *manages* elements of the contract meaning that the contractor has to live with the project once it has been built. When the private sector is risking finance, there are far greater incentives to get everything right, especially as building contractors now have to consider the running and maintenance costs.

The model structure of a PFI contains an element known as a Special Purpose Vehicle (SPV) (Figure 11.1). The SPV is the project company established by the sponsors, who have the sole purpose of delivering a project. The SPV draws on requirements set by the public-sector client and the private-company contractor to ensure there is a successful building for use in public service.

The value for money that a PFI-funded project brings is particularly pertinent for those involved in building procurement. Key elements in whether value for money is reached are those such as competition and the cost of construction, cost overruns, risk, and if there is in fact any evidence of value for money. Recent governments in the UK and EU have seen competition for public services as central to public procurement, to guard against corruption and secure value for money.

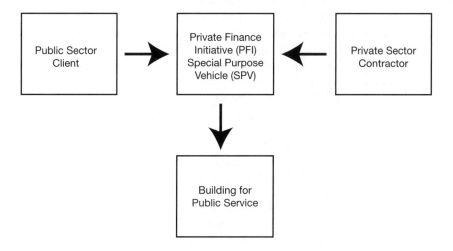

11.1 Structure of PFI procurement

Examples of savings through competition are the Bridgend and Fazakerley prison projects that received from a tender in the *Official Journal of the European Union* (*OJEU*) over sixty expressions of interest, five of which were invited to submit bids. The Bridgend project selected a consortium (Securicor/Costain) that was over £50 million less than that of a similar public-sector comparator (PSC). In a similar project, the Fazakerley prison was not selected only on the basis of quality, as the prison project's procurement procedures recognised that the same contractor (Securicor/Costain) could not simultaneously build both projects to sufficient quality. The tender went to the second-lowest bidder in this instance (Group 4/Tarmac consortium). It is also noted in this example that Fazakerley prison also made £3.4 million savings through reducing commissioning and construction costs, and the prison opened five months ahead of schedule, which made further resource cost savings.

Transfer of risk is a contributor to gaining value for money in building procurement and has been key to the use of PFI. Risks can be general but are often more specific and dependent on the project in question. The risk involved in a school project will differ from that for a highway project, for instance. The important aspect relating to value for money in the public sector will be that for each project there is sufficient risk transferred to the private sector. The level of risk transferred can be subdivided into the transfer of disposal risk and external finance risk. The amount of disposal risk is the level of assets that are surplus to public-service demand at the point of disposal. External financial risk will be determined by the ability to raise sufficient funds on the open market. This will be intrinsically linked to the changes in interest rates, as a greater than expected rise in interest rates will be detrimental to lending repayments. Transfer of risk also means that there is a possibility of the project failing, which means that transfer should be given to those private organisations that are able to manage risk effectively.

Evidence of value for money in procurement using PFI in public-sector projects has previously been difficult to obtain. The NAO has started to publish a series of reports that record PFI deals, although there is relatively little evidence on the performance over the long term of individual PFI projects (NAO 2010). A general saving figure that can be drawn on is the difference between the present values of public-sector comparators (PSCs) and the present value of the winning bidders. Using an NAO sample, it was calculated that the total present value of the PSCs of seven PFI projects was over £4.6 billion, compared to the total present value of the winning bidders of just over £3.7 billion – indicating a saving of £0.9bn or 20 per cent (Arthur Anderson and LSE Enterprise 2000). It can also be inferred that if the sample was representative of all PFI projects signed up in 2001, valued at £22 billion, there would be a 20 per cent saving in the region of £4.4 billion using this PFI procurement approach.

A great deal of gloom has been cast more recently over the future of partnering using PFI. There have been repeated warnings by government that construction's capacity to take on PFI projects will dwindle as the costs of bidding begin to outweigh the likely rewards. It has also been pointed out that contractors tend to develop niche markets in, say, health and/or education and shun those sectors that are troubled by political 'shilly-shallying' (Chevin 2002). To compound these issues even further, a recent Directive from the European Commission suggests that at least five consortiums need to compete in the final bid for major projects. This in turn could present yet another hurdle in developing and extending PFI and prime-contracting arrangements.

This highlights the economies of scale in partnering, where it is easier for large firms to gain advantage and benefit further from partnering arrangements. Small firms generally lack the resources and expertise to participate in the lengthy bidding process involved in PFI contracts.

In construction where PFI is emerging, competitive market conditions do not prevail. Many firms lack the necessary resources to understand the complex legal information that is inevitably associated with these forms of procurement. Transaction costs are prohibitively high, with architects', lawyers' and accountants' fees needing to be met by all the participating parties. As a result, it is unusual for more than three or four consortium groups to find sufficient resources to engage in the tedious, lengthy and detailed bidding processes involved. Indeed, PFI bidding costs can commonly exceed £1 million per project. The firms that are able to take on such large-scale operations and risks are few and far between, and it is a common concern that partnering arrangements often exclude the smaller contractors.

Much of the scepticism around the use of PFI for building procurement can be seen to centre on the issues of high bidding costs, the cost of borrowing, European procurement Directives and poor negotiation when refinancing PFI contracts. High bidding costs are often put down to high 'front-loaded' costs that are involved in organising and preparing a tender for a PFI project. The cost of borrowing may in fact be more costly from private funds in comparison to sources of finance that can be accessed by the public sector. PFI funds from the private sector may enable a greater variety of financial borrowing options but may not necessarily be cheaper than public sources such as the National Loans Fund (NLF), as determined by the Treasury. The NLF sources of finance are backed by tax revenues and are risk-free and hence a cheaper way of acquiring funds.

Issues regarding PFI as a procurement option with respect to European Commission procurement Directives involve the size of bid and the loss of control in selecting a preferred bid. Contracts above a certain threshold are put out to competitive tender in the *OJEU*. UK projects are selected on the basis of the offer having the most economically advantageous bid. Proposals for a new EU Directive on public procurement may take away this right and insist that at least three bidders will be required to continue negotiations until the finalised PFI contract is signed. Legislation on public procurement contracts in the EU can be found at the European Commission website that documents such Directives – this includes guidance on setting up institutionalised public–private partnerships (European Commission 2010).

Weak negotiation skills in drawing up PFI contracts are another issue that questions the value for money when using PFI for procurement. For instance, some contracts allow a refinancing option for the private sector part-way through a project. It can be argued that this refinancing option should integrate an option for the public-sector client to be reimbursed. If not, the private client will take some reward for any positive changes in the external financial market. For public-sector clients to prosper from refinancing at stages during the project it will take contract negotiation and legal skills on the part of public managers.

The longevity of PFI for public-service building contracts has shown signs of waning in recent years. Construction statistics have shown that between 2007 and 2008 the percentage of building contracts ordered and financed by PFI, as defined by demand from new orders, fell from 5 per cent to 3 per cent (ONS 2009). Movements in the same period saw a percentage change increase for public-sector demand for construction up by 10 per cent to produce a total public-sector demand from new orders at 34 per cent. This contrasted with a decrease of 7 per cent in private-sector new orders for building over the same period between 2007 and 2008; this resulted in 63 per cent of new orders being demanded by the private sector for private use (CSA 2009). This fall in private-sector investment in both private and public building is in part reflective of the economic recession. Falls in GDP often create more volatile and initial ripple effects in the private sector. Future uplifts in the economy may reverse the rising proportion of public-sector building projects. A growth of PFI-funded projects may return if there are limited public funds to invest in building public services. The availability of public funds will be further stretched during times where there is pressure to reduce a country's PSBR deficits.

As for the future of PFI, the borrowing requirement rise to £1.2 trillion in 2009 means that the PFI is more likely to remain as a procurement tool but in a more complex form. It is argued that complexities in using PFI are currently more challenging because the financial sector is now weak compared to previous periods. For instance, the financial sector remained relatively intact during the last recession. This time round PFI can be seen as the principal villain (in terms of toxic lending) and victim (in terms of failed initiatives) of contemporary economic development practices. The rather bleak view that the collapse of private credit has dragged the private finance initiative down with it has been challenged. For instance, it is argued that government building programmes are so important to social and economic imperatives that a way will be found to build public infrastructure in the future (*Public Private Finance* 2009a).

During a time of restricted credit and finance, deals may more likely take the form of Public–Private Partnerships (PPPs) rather than PFIs, especially if the lack of liquidity is forcing the government to look at using PPPs far more broadly. There appears to be no halt in procurement from the private sector in building public-sector projects. For instance, private healthcare is back on the agenda, there is rising council outsourcing and there are talks of private firms running education authorities. These public–private partnerships are now considered.

11.4 Public–Private Partnerships (PPP)

A Public–Private Partnership (PPP) is a public service or private business venture that is funded and operated through a partnership between the public sector (either central or local government) and one or more private-sector companies. PPPs are recognised as a key element in the government's strategy for delivering modern, high-quality public services and promoting the UK's competitiveness (HM Treasury 2000). The use of PFI is one element of PPP business structures and partnership arrangements. Others include joint ventures, outsourcing and the sale of equity stakes in state-owned businesses. This introduction of private-sector ownership into state-owned businesses can have a full range of possible structures. The structure could for instance be brought into existence through floatation or the introduction of a strategic partner. Either public- or private-sector interests can hold major or minor stakes in the PPP. The importance of PPPs to building procurement is the amount of funds and policy support directed at the construction industry in using this model of partnership.

PPPs are therefore a particular type of contractual arrangement between the public sector and private-sector firms. They give the private sector a greater role in financing, building and maintaining public-sector facilities, although the government retains a stake in the PPP company. In contrast to the PFI, under public–private partnership arrangements the government is not liable to a fixed stream of annual payments. PPP is therefore an arrangement that can be financed via both public-sector and private-company sources. For instance, a partnership contract can be drawn up that recognises agreed government funding and private-developer contributions to a project.

With the introduction of the private sector in PPPs, advantages for the public interest are maintained if fundamental government roles are made responsible and held accountable. Fundamental roles for the government include being the principal decision-maker between different competing objectives. This then allows the government to retain authority as to whether the objectives are delivered to the standards required. Most importantly the fundamental government decision-making should in theory ensure that the wider public interests are safeguarded. Public-interest issues would be those such as putting in place regulatory bodies that remain in the public sector, maintaining safety standards and ensuring that any monopoly power is not abused.

To emphasise how PPPs have radically changed procurement in the public sector, examples of conventional approaches are contrasted against current government visions. An example of overspend and delay in the public sector is Guy's Hospital, London, which was completed three years after completion date and with an initial cost estimate of £36 million that ended as a final cost of £160 million. In transport, the Jubilee Line Underground extension was delayed by almost two years and an initial cost estimate of £2.1 billion finally came in at £3.5 billion. This £1.4 billion overspend further demonstrates the need for either more accurate projections or improved procurement methods in such projects. This is not to say that initial budgets are initially kept artificially low, but are often projected low in the first instance to ensure that projects using public money win the bid and get the go-ahead.

With public criticism of public-sector overspend, the government asked Peter Gershon, then managing director of GEC Marconi, to carry out a wide-ranging review of civil procurement. This commission sought to modernise public-sector procurement and ensure that some of the principles established through PFI should become standard practice within the public sector and PPPs. This led to the creation of the Office of Government Commerce, as has previously been discussed, to which Gershon was appointed Chief Executive in 2000. Key recommendations in rethinking procurement included the introduction of equitable due diligence in public procurement, as is carried out in private business. Moreover, there is a call to involve rigorous expert scrutiny not just at the point of purchase, but also throughout the life cycle of public-sector procurement projects. The recommendation to incentivise procurement in the public sector was especially significant. It was particularly stressed that procurement should not just be a 'sell off' by emphasising that public-sector parties benefit from the innovative approaches developed by private-sector partners in PFI deals.

The complexity and intricacies of integrating private-sector risk into procurement has meant that more sophisticated PPP design structures have come into existence. As an exceptional case the Channel Tunnel Rail Link (CTRL) PFI project devised a novel deal structure to rescue the project. This involved the government guaranteeing bonds that were issued by the London Continental Railway. This type of financing was rare because of the exceptional nature of the project and the low risk that the bonds would be called for payment. In essence there would only be one Channel tunnel and its success was as good as guaranteed. The raising of finance through bonds issued by state-owned businesses or bonds guaranteed by government does not offer best value. It is more cost-effective for governments to issue (non-guaranteed) gilt-edged securities (or gilts) to directly finance projects as they offer less guarantee and more risk but as a result can offer greater reward – therefore generating greater initial funds to finance the project. Increases in the PSBR are avoided through the direct issue of government gilts into the financial market, rather than state-owned businesses issuing bonds for the direct financing of its PPP projects.

More generally, partnering in any form of procurement is an attractive idea and in economic terms it helps to eliminate inefficiency as costs per unit of output are reduced. Leibenstein's (1973) concept of x-inefficiency can be applied here. This is where the decisions of an individual public-sector authority or an individual contractor are less likely to maximise the value of output, particularly when applied to decisions on what to build, how to build it and how long to spend on the project. Partnering also improves the dynamics of the market by putting the client more in control and improving the flow of information between the participants. Partnerships can also provide the platform to offer greater incentives to complete the contract on time, on budget and to the expected quality. Some of the benefits of partnering are summarised in Table 11.6, suggesting how the construction industry in the UK could become more efficient.

Contemporary issues surrounding the partnership industry involve the reduction in credit available to finance projects and the subsequent programme delays and new deals being signed. More than sixty PFI deals had been signed each year since 1998, with 67 deals signed in 2007. This consistent number of deals took a fall in direct correlation to the credit crisis, with a slump to half as many deals on the previous year at just 34 (*Public Private Finance* 2009a). The key projects that stemmed from this slump were those involved in health and education building programmes. There were just seven health deals signed in 2009, down from twenty in 2007 as polyclinic building failed to compensate for the truncated hospital-building programme. Furthermore, there were just eight BSF deals signed in 2009, down from ten in 2007. As well as a reduction in deals, credit has had a negative hit on partnerships in other ways. In privately financing partnerships, PFI bond markets closure added to PFI problems as monoline insurance companies solely for bonds, which had insured the bonds, were downgraded. As a result, banks no longer felt able to take on large chunks of debt, and the cost of finance rose sharply. In the next chapter, discussion is set against this backdrop of a shrinking public purse and a tight financial market, and focus is given to the critical procurement decisions to be made by both public and private stakeholder partners.

Table 11.6 The benefits of partnering

1 Reduced need for costly design changes.
2 Increased opportunities to replicate good practice learned on previous projects.
3 Avoids adversarial relationship between client and contractors.
4 Contractors have good incentives to deliver on time, on budget and to a high standard.
5 The liaison of clients and contractors should improve the overall efficiency of the building particularly in terms of its operation and maintenance.
6 It should be possible to drive out inefficiency and waste from the construction process.

Source: National Audit Office 2001: 31.

11.5   Summary and tutorial questions

11.5.1 Summary

Main features of public-sector projects

- All projects that are under ownership by central and local government.
- Most companies bidding for public-sector work will encounter a formal procurement process.
- European Directive 2004/18/EC came into effect in 2006 to drive pan-European fairness, value for money, transparency and accountability.
- Most projects out for tender have a pre-qualification element and a structured system of procurement.

The Office for Government Commerce (OGC) Gateway process

- OGC is involved in the procurement of goods and services in the building sector.
- OGC promotes several mechanisms with an aim to professionalise procurement and make it more efficient and effective.
- The development of the Gateway process examines programmes and projects at key decision points in their life cycle.
- A major element of the Gateway process is through the use of reviews such as peer reviews that are made during the life of a project.

The Private Finance Initiative (PFI)

- PFI is a form of procurement to encourage private investment in public-sector projects.
- Introduced in 1992 by the Conservatives with continued support from 1997 by New Labour with high-profile projects such as Building Schools for the Future (BSF).
- Set against a backdrop of continued government objectives of decentralisation and liberalisation of markets.
- Also part of wider processes of increased downsizing, outsourcing and developments in information technology.
- Under PFI the purchaser does not immediately acquire the assets but puts in place a contract to buy services from the supplier. The assets may remain with the supplier for up to thirty or forty years.
- Private investment may include the design, building, financing and operations as part of the public–private contract to develop the asset.
- A Special Purpose Vehicle (SPV) draws on requirements set by the public-sector client and the private-company contractor to ensure that there is a successful building for use in public service.
- Value for money in PFI is reached depending on the level of competition, cost overruns, and risk.
- Doubts around the use of PFI for building procurement can be seen to centre on high bidding costs, the cost of borrowing, European procurement directives and poor negotiation when refinancing PFI contracts.

- Capacity to take on PFI projects may dwindle if credit restriction continues and asset values stagnate.

Public–Private Partnerships (PPP)

- PPP is a public-service or private-business venture that is funded and operated through a partnership between the public sector and private-sector companies.
- In contrast to the PFI, under PPP arrangements the government is not liable to a fixed stream of annual payments.
- PPP is an arrangement that can be financed via both public-sector and private-company sources.
- Benefits are argued to be a reduction of procurement costs and a reduction in project delays
- The complexity and intricacies of integrating private-sector risk into procurement has meant that more sophisticated PPP design structures have come into existence such as introducing specific project bonds (as for example in the case of the Channel tunnel).
- Increasing the PSBR is avoided through the direct issue of government gilts into the financial market, rather than state-owned businesses issuing bonds for the direct financing of its PPP projects.

11.5.2 Tutorial questions

Case study: the procurement of works and services for the design, construction, operation and maintenance of a built facility on behalf of a government client

The project involved the design, operation and maintenance of a facility together with 15 different service lines related to the operational requirement. The capital expenditure of the project is about £70m and the net present cost of the operation of the facility over a 25-year life is about £350m.

The bid assessment was designed to support the selection of the most economically advantageous offer. The overall objective of the evaluation process was to support a consistent process, ensure transparency and objectivity, determine the most suitable bidder to proceed to a final negotiation and, through the formal assessment system, provide feedback to all bidders.

The multidisciplinary appraisal team represented 20 disciplines, each with its own set of stakeholders and deliverables. Assessment and stakeholder management was a key part of the project management service offered. An evaluation process guide was prepared, setting out the basis of the evaluation, the process, team, programme and the selection strategy. This document became the key control document for the appraisal process.

An invitation to tender (ITT) was sent to suppliers who had pre-qualified on the basis of an assessment of technical capability, financial stability, contractual compliance and commercial competence. The ITT package included a high degree of contact with the procuring authority to ensure that the tenderers understood the requirements, together with specific workshops designed to enable bidders to discuss their technical, legal and financial proposals. After submission of the bids, the tenderers would be given a series of opportunities to present their proposals during the evaluation period.

In evaluating the bids, the key criteria identified by the authority were whole-life costs, design (of the service delivery, as well as the physical facility), quality, delivery, management, innovation and supplier-specific risk. While cost was evaluated on the basis of hard cost, other aspects of the bid were assessed against weighted criteria of design, quality, delivery, management and innovation. Design of the overall operational solution received the highest weighting of 30%.

A further concern of the assessment team was the relationship of the construction stakeholder and the rest of the consortium, to ensure that the proposed solution was properly managed, integrated and 'operationally led', and that the capital works would not have a negative effect on whole-life value. Scoring of each technical criterion was based on a scale of 0–5, where 0 meant no compliance and 5 was a premium representing significant and relevant over-compliance.

Tenderers were given a detailed list of deliverables with guidance as to what was required in each section of the tender submission. Sections of each tender submission were given to specific groups within the evaluation team. A management structure and programme was put in place to provide clear direction to the team. This included task descriptions for each evaluator, an assessment panel to adjudicate queries from the assessment teams, and a group of lead evaluators. Processes were also put in place to deal with tender queries, variant bids, bid-specific risks and so on.

The overall period of assessment, including the preparation of a report, took a little over three months, including the completion of a report recommending that the client should finalise terms with the preferred tenderer.

Source: *Building*, issue 47 (2006)

1   Using the case study involving a new-build facility on behalf of the government, discuss which of the nine success factors for a successful evaluation system have been used.
2   Using BSF as an example, bullet-point review outcomes to consider in the OGC Gateway process review stage model.
3   What are the differences and similarities between PFI and PPP?
4   Outline the advantages and disadvantages to the public sector when using PFI to fund construction projects.
5   Explain whether partnering and the use of PPP always creates a win–win situation.

12 The selection of building procurement systems

12.1 Introduction

In the previous chapters, the various procurement systems have been identified and described, and it is from this wide range of means of procuring the design, construction and other aspects of the project that the client has to select the most appropriate method to ensure that their needs and objectives are met.

In order to understand how such selections should be and are made, it is first necessary to recognise the principles of decision-making and choice in general. The *Concise Oxford Dictionary* defines 'decision' as a 'settlement (of question, etc.), conclusion, formal judgement; making up one's mind, resolve; resoluteness, decided character', which effectively boils down to a simpler and more straightforward definition: 'The process of choosing one action from a number of alternatives.'

However, the act of reaching a decision (decision *taking*), no matter how the act itself is defined, is only the final stage in the more dynamic process of decision *making*. When definitions of the whole process are examined, a common theme can be extracted of a dynamic process which starts by the identification of need and culminates in the act of making a choice between alternative means of satisfying such need(s).

The examination of decision-making that now follows is therefore carried out by looking at the process in its entirety rather than just the final act of reaching a decision.

12.2 Theoretical decision-making and choice

12.2.1 The decision-making process

Although decision-making can take many different forms and can be carried out by individuals, groups or organisations, the process and stages within it remain constant and have certain common features. The decision-making process is relatively straightforward and is made up of seven sequential steps:

1   set the objectives;
2   search for alternatives;
3   compare and evaluate alternatives;
4   choose among alternatives;
5   decide (implement the decision);
6   follow up and take corrective action if necessary;
7   revise and update objectives.

The process in this model is circular, dynamic and iterative, with each stage constantly referring back to the previous one. Therefore, the common elements of simple, uncomplicated decision-making are:

- Decision-making is a circular/continuous activity, although in the case of the selection of a procurement system for a single project the process normally lasts only for the duration of the project, although decisions made regarding design and many other aspects of the project will have an effect on the management of the facility for the whole of its life and will thus be part of a continuous process.
- Most, if not all, organisations have a continuing requirement to establish objectives in order to meet their ever-changing needs, and, in the process of attempting to meet these objectives effectively, a variety of decisions must be made.

The decision that concerns us here is the selection of the most appropriate procurement system to satisfy the client's various needs related to a specific project or programme of projects.

- To assist in making such decisions, the use of a framework which enables the organisation to move through the decision-making process in an orderly and efficient manner must be helpful.
- The framework will normally consist of a series of simple steps, whereby the problem is identified and diagnosed; objectives are then set and constraints considered; alternative solutions/courses of action are identified; a choice is made; and the consequent decision implemented.

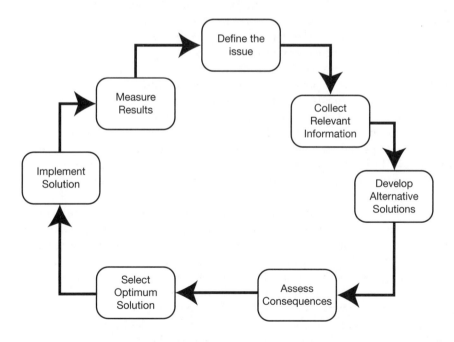

12.1 Standard model of decision-making

- Once execution has taken place, or is continuing, the activity is reviewed in order to establish whether the original problem identification, investigation and choice of solution were correct or whether further remedial action is necessary.

In the case of procurement system selection, once the choice has been made and implemented it is past the point of no return and any attempt to change the method of procuring the design and construction will be likely to be too costly and result in delays and therefore extra costs to the project. This situation reinforces the need to delay the selection until the latest possible time and also ensure that every effort has been made to make sure the choice is correct in the circumstances.

While the decision-process framework outlined above would probably be seen as an acceptable, if simplistic, view of the theoretical decision-making process, there are in fact three different decision-making models:

1   rational decision-making
2   behavioural decision-making
3   political decision-making.

12.2.2 Rational decision-making

The rational approach was developed by mathematicians, statisticians and economists and is encapsulated in subjective expected utility theory, which is a very grand way of saying that the decisions are made based on what the client (decision owner) expects to achieve or get out of the process.

The theory revolves around the concept of rationality (i.e. they know what they are doing) and assumes that the decision-maker not only knows exactly what they want but has an identifiable set of alternatives to choose from and is aware of their outcomes; is able to choose the alternative that gives the best value; and is able to ignore any constraints within their environment.

Examination of these criteria soon reveals that, while the rational approach is valid where the decision-maker is in possession of perfect information and is able to meet all of the identified requirements, very few individuals or organisations operating in the real world are ever in such a state of certainty when making decisions. Usually, decision-makers are in possession of only part of the necessary information needed to examine all the alternative solutions and determine their consequences. Often, organisations have difficulty in accurately identifying the cause of the problem that is compelling them to enter into the decision-making process and are unable to produce even a simple list of alternative means of solving the dilemma.

Certainly, in the case of clients of the construction industry, their 'problem' is usually relatively easily identified – they require some form of facility that will enable them to provide a service or a product – but in most cases they are not in possession of all of the necessary information to allow them to make a rational decision as to how to achieve their objectives. In an attempt to deal with these practical difficulties, an alternative theory of bounded rationality, or satisficing, has been developed which asserts that decision-makers are hardly ever in a position to be fully rational, i.e. they operate under conditions of uncertainty and, in reality, reduce the complexity of decision-making by constraining the process of developing alternatives and processing information.

However, this approach additionally assumes a perfectly unitary decision-maker, with information, knowledge and judgement shared among the responsible members of the organisation and with everyone agreed about the goals, values and perceptions; a situation which rarely, if ever, exists in any corporate establishment.

12.2.3 Behavioural decision theory

This is essentially an expansion of the above 'bounded rationality theory' but with further ideas of the mechanics of the decision-making process used in large bureaucratic organisations operating under uncertainty in imperfect markets.

This theory suggests that alternative solutions are not necessarily compared with each other, but are examined sequentially and accepted, or rejected, on the basis of target aspirations or their consequences. Where, during the decision-making process, there are any internal conflicts of interest, these are controlled by buffering, or isolating, inconsistent demands from the interested parties for lengthy periods. Rule-following, more than the calculation of the consequences of implementing alternative actions or solutions, determines most decisions, and decision-makers adopt strategies which avoid the ambiguity that exists within much of the decision-making process.

Although incorporating most of the substance of decision-making, the behavioural theory does not address the possibility that underlying any collective decision is a process of bargaining in which choices are determined by the resources committed by each member of the organisation to the achievement of some satisfactory solution.

12.2.4 The political bargaining model

This model suggests that a process of bargaining between participants enables a final decision to be made which has the general support and accommodates the interests of all the participants.

While both the rational and behavioural approaches still have their advocates, the decision-making process is often an amalgam of all three models, although agreement on the relative importance of each is unlikely to be reached.

Organisations operate in an environment in which external sources such as the country's political institutions, culture, traditions, laws, etc. exert pressures of varying kinds upon the decision-maker. Where choices and decisions are made without cognisance of any element of the environment, either they are unlikely to be capable of being implemented or problems will arise down the line. The organisational decision-maker is also affected by the economic systems within which the organisation operates, i.e. the economic environment, which includes customers, competitors, the supply chain, associated industrial organisations as well as local and central government.

The effect of the economic environment is often overlooked when choosing a procurement system and tendering strategy. However, this point was certainly not overlooked in June 2009, when the British Airports Authority (BAA) decided to revert to competitive tendering from its previous stance of framework arrangements (see case study in Chapter 10). They felt that they could obtain more competitive bids in the economic climate of the time. This possibly also shows that the philosophy of frameworking was never fully engaged.

When the order books of contractors, subcontractors and suppliers are full, they will only be tempted to submit bids when the method of procurement is attractive, risks are minimised and it is likely that the successful tenderer will achieve an acceptable level of profitability on completion of the project. Again, great care must be taken here as the government's Office of Fair Trading in the UK considers that the practice of 'cover pricing' (putting in a high tender with the intention of *not* winning the work) is a form of collusion and therefore illegal. In late 2009 very large fines were imposed on many contractors found to have operated the practice.

The actions of government are particularly relevant to the decision-making process. In order to ensure that the country can be fairly and effectively governed to the general benefit of its subjects and citizens, that the economic system is properly regulated and that relationships between parties are equitable, legislation is required. Unfortunately, the laws and acts stemming from this need place a set of restraints upon the decision-maker within which choices must still be made in order to meet the organisation's objectives.

The decision-making process is also influenced by the social fabric of our world; the forces of social change and the plethora of social problems exert continuing and increasing pressure on corporate decision-makers. Irrespective of other influences that may affect the environment in which any decision has to be made, these problems increase the level of inherent uncertainty and risk already present. The economic system is not only concerned with what, and how many, goods and services will be produced but also with how, when and where these will be exchanged.

Therefore, in theory at least, the decision-maker – in addition to being in a constant state of uncertainty – must deal with the idiosyncrasies of their own organisation's structure and politics and at the same time take into account the economic, social and legal consequences of their eventual decisions.

12.3 Decision-making in the real world

Let us now look at how decisions are actually made in industrial and commercial organisations and, where appropriate, compare theory and practice.

12.3.1 Models of procurement route selection

As we have seen in the classic ideal, decision-making is seen as a series of logical steps with clearly perceived goals and with the ability to conduct comprehensive searches and obtain complete information; in other words, an orderly, well-disciplined process.

When dealing with the actual making of what are often called *programmed* or *structured* decisions, this view has some validity, as such decisions are often routine in nature. Established organisational procedures, often based upon past experience, exist to deal with them and the solution is often obtained by a logical application of these procedures.

In construction and development, such decisions are regularly made, e.g. when carrying out development appraisals and planning the implementation of simple repetitious projects where there is ample information available on the technical and managerial aspects and where the client organisation, and its advisers, are familiar with the type of project and are able to determine the project's parameters in the knowledge that few, if any, changes to the brief will be made through the duration of the work.

However, when describing the making of strategic decisions within organisations, these are non-programmable *unstructured* processes which may require judgement, experience, etc. Such decisions, when they are made within organisations, are for the most part incapable of being programmed owing to their complexity, ambiguity, susceptibility to external influences and lengthy gestation period. Strategic decisions are associated with the long-term management of the organisation and concern the goals to be pursued and the strategies needed to achieve them.

A strategic decision has a number of characteristics that differentiate it from other lesser decisions. It will be comparatively novel, being less common and more non-routine than most; there will be fewer precedents for it, but it may itself set precedents for future decisions; it will commit substantial resources; it is often organisation-wide in its consequences and triggers other lesser decisions; and is thus atypical, important and all-pervading.

A long-term programme of new-build construction and the implementation of most major projects, and indeed the selection of the most appropriate procurement method to use for the implementation of such developments, exhibit many of these character-istics and can therefore be treated as strategic decisions.

Strategic decision-making is a disorderly process and in reality does not conform to the ideal phases of decision theory, with stages being skipped, taken out of order, taken concurrently and generally appearing to be ill-defined. Notwithstanding all of this, decision-making is a fundamental part of organisational life and as such needs to be managed.

The majority of strategic decisions appear to arise from deliberate managerial strat-egies which reflect the organisation's aims and objectives, although difficulty is often experienced in determining what the problem/opportunity actually is and/or setting objectives to deal with it. Most of the action in the decision-making process takes place when alternative solutions and/or courses of action are being identified and developed. Conflict can occur during this lengthy phase as information and opinions are obtained and as alternatives identified and considered.

Uncertainty is present at the evaluation stage in great abundance as a result of the lack of knowledge of the consequences of the various alternatives being considered. Under these circumstances, decision-makers, while using as much data, experience and beliefs as they can muster, often fill some of the gaps with hope and expectations.

The choice process, which should consist of the evaluation of alternatives, screening by the use of secondary constraints and determining which alternative enables primary goals to be best met, is distorted by a surfeit of information, bias, change, politics and internal and external interest. More difficulties are experienced during the authori-sation of the choice than in the evaluation phase, as the decision often needs to be considered within a limited timescale and in the light of other contemporary decisions, with the authorisers often lacking sufficient detailed knowledge of the proposal and sometimes being subject to external pressures.

The decision-making process in practice is therefore not the classic model of problem-solving following a logical sequence beginning with the problem being defined and continuing sedately and smoothly through scanning information, assessing alterna-tives and choosing a solution. In the real world the process usually begins with very little understanding of the decision situation, a vague notion of possible solutions and very little idea of how to evaluate them and choose the most suitable solution. It is iterative, recursive, discontinuous and ever-changing; it also takes place over a consid-erable period of time and derives from a combination of its organisational setting and the problems and interests associated with the matter of choice.

The construction industry is a fragmented, diverse industry responsible for the design and construction of bespoke and often extremely complex, expensive and lengthy projects carried out in a particular time and location by means of what is usually a temporary management organisation consisting of groups of highly qualified and individualistic professionals who are frequently unknown to each other prior to project commencement. Not a particularly auspicious way of guaranteeing a successful outcome, but it is worth noting that the main stadium for the London 2012 Olympic Games was given to the consortium of companies who had very successfully built the Emirates stadium for Arsenal Football Club, mainly because of the good working relationships fostered on that project.

The environment in which projects are implemented is thus far removed from the relative stability, permanence and sense of company loyalty which usually exists in industrial and commercial client organisations. It is generally accepted that, as a prerequisite to the implementation of any project, the client – particularly the inexperienced client – should have an understanding of the workings of the construction industry, its procedures and the characteristics of temporary management organisations, as well as having some overall knowledge of all of the available procurement systems.

The steps that a client should take to ensure the best possible success of a project are:

1   appoint an in-house project 'owner' (sometimes known as a 'project sponsor');
2   appoint, if required, a principal adviser;
3   carefully define the project requirements;
4   realistically determine the project timing;
5   select an appropriate procurement path;
6   consider the organisations that will be employed;
7   professionally appraise the proposed site/building before making a commitment.

Assuming that the first two steps have been taken and that the client has an understanding of the construction industry, the third and fourth steps – defining the project requirements and determining the project timing – lead us back to the fundamental need to establish a clear and comprehensive brief. This should contain not only the aesthetic, technical and performance criteria for the project but also, of equal importance, the primary and secondary objectives in terms of functionality, quality, time and cost and the political, cultural and economic environment within which the project will be implemented.

However, the definition of the project requirements needs to be broken down into two stages: first, a review of the client's needs, which requires that the construction industry must be capable of understanding organisations as well as managing design and construction, and, second, a definition of the means by which these needs will be achieved.

The UK government's Office of Government Commerce (OGC) publishes guidance notes on procurement and contract strategies which provide specific help in defining the means by which the selection of an appropriate procurement system can be achieved. The factors that influence the procurement strategy are:

1   The project objectives, for example, to provide office space for x people to deliver a specific service.
2   Constraints, such as budget and funding; the time frame in which the building is to be delivered; exit strategy.
3   Cultural factors, such as considerations about the workspace environment that will best support the way people work.
4   Risks, such as late completion of the facility, innovative use of materials.
5   The client's capability to manage a project of this type.
6   The length of operational service required from the facility.

The procurement process then follows the stages described in Table 12.1.

In terms of the factors which might influence the choice of procurement route, JCT Practice Note 5 *Deciding on the appropriate form of Main Contract* lists the following:

Table 12.1 The procurement process

Procurement stage	Gateway review	Key procurement tasks to each gate
Establish business need	Gate 0: Strategic assessment	Identify high-level options for meeting the business need
Develop business case	Gate 1: Business justification	Produce high-level business case (Strategic Outline Case) and detailed options appraisal
Develop procurement strategy	Gate 2: Procurement strategy	Produce Outline Business Case; determine procurement route (including contract strategy); produce output-based specification and criteria for selection and award; *OJEU* advertisement if required
Competitive procurement	Gate 3: Investment decision	Competitive tendering (where there is no existing arrangement with a supply team) leading to contract award for integrated supply team; full business case
Award and implement contract; outline design	Decision point 1: Outline design	Following approval of outline design, proceed to detailed design
Detailed design	Decision point 2: Detailed design	Following approval of detailed design, proceed to construction
Take delivery of facility; settle final account/start unitary payments	Gate 4: Readiness for service	Commissioning of facility; handover to contract management where applicable
Manage contract for services, where applicable	Gate 5: Benefits evaluation	Post-implementation review, to confirm achievement of business benefits as the justification for investment in the facility

Source: UK Office of Government Commerce.

- *The nature of the project*
  For example – is this a completely new detached building; an extension to an existing building; a refurbishment job; restoration of an historic structure; reinstatement after fire damage or neglect; a repair and maintenance programme involving many buildings?
- *The scope of the works*
  For example – is there something unusual about the size, complexity or location of the works; are there site problems of access, storage or movement; does the work involve the basic trades and skills of the industry; does an innovative design demand sophisticated construction methods; is there specialist subcontractor's work with a design content; is there a high content of specialist engineering installations; is this a single construction operation, phased work, or part of a term programme?
- *Measure of control by the client*
  For example – should design be wholly in the hands of the client's consultants; can some detail design be placed as a contractor's responsibility; should there be provision for design by specialist subcontractors; to what extent does the client wish to control selection of specialist subcontractors; what measure of control will the client wish to exert over materials and workmanship; how much reliance can be placed on performance specified requirements?
- *Accountability*
  For example – does the client aim for single point responsibility; is it the intention to appoint a project manager or client's representative; where is responsibility intended to lie for specific matters – with consultants, contractor, specialist subcontractors?
- *Appointment of a contractor*
  For example – is this to be by negotiation or by competitive tendering; is the contractor to be appointed to carry out construction work only; is the contractor to have some responsibility for design; is the contractor to be appointed early to undertake primarily a management role?
- *Certainty of final cost*
  For example – is a lump sum contract preferred; will it be a fixed price or with fluctuations; do the circumstances dictate remeasurement and an ascertained final sum; must all tenders be on a competitive basis?
- *Start and completion times*
  For example – is this to be 'fast track' with the shortest overall programme a priority; is an early start date desirable; will there be adequate time to prepare full information for tendering purposes; do circumstances dictate a specific completion date; can the contractor be provided with exclusive possession right from the start?
- *Restrictions*
  For example – does the site raise security problems or problems in relation to surrounding property such as access or noise; are there restrictions on working hours; will the building be still in operation and occupied during the course of the works; is the work to be phased; is there a specific requirement concerning the sequence of operations?

- *Changes during construction*
  For example – is there a likelihood of design changes during the course of the works; can the contract satisfactorily accommodate variations and the valuing of such work; to what extent might approximate quantities or provisional sums be required?
- *Assessment of risks*
  For example – is this to be a contract with the lowest possible risk to the client overall; what are the priorities in apportioning the risks concerning cost, time, and quality or performance; where are the speculative risks intended to lie?
- *Building relationships with the supply chain*
  For example – is a long term relationship with a supplier or the supply chain required so as to provide continuous improvement?

Source: Joint Contracts Tribunal 2008 (© JCT)

Furthermore, in terms of the appropriateness of the various procurement routes to the above client criteria, Table 12.2 gives useful guidelines.

Table 12.2 Appropriateness of procurement routes to client objectives

Project objectives parameter	Objectives	Appropriateness of contract strategy in meeting project objectives				
		Traditional	Construction management	Management contracting	Design and manage	Design and build
Timing	Early completion	✗	✓	✓	✓	✓
Cost	Price certainty before construction start	✓	✗	✗	✗	✓
Quality	Prestige level in design and construction	✓	✓	✓	✗	✗
Variations	Avoid prohibitive costs of change	✓	✓	✓	✓	✗
Complexity	Technically advanced/ highly complex building	✗	✓	✓	✗	✗
Responsibility	Single contractual link for project execution	✗	✗	✗	✓	✓
Professional responsibility	Need for design team to report to sponsor	✓	✓	✓	✗	✗
Risk avoidance	Desire to transfer complete risk	✗	✗	✗	✗	✓
Damage recovery	Ability to recover costs direct from the contractor	✓	✗	✓	✓	✓
Buildability	Contractor input to economic construction to benefit the department	✗	✓	✓	✓	✗

The table is clearly for guidance only. Generally, the appropriateness of the procurement route is never as clear-cut as indicated and it is important that the project manager should advise the client's project sponsor on this and give advice on the fine tuning of the procurement decision.

The procurement selection process has become increasingly complex in the modern era, mainly as a result of the continuing proliferation of different methods of procuring building projects, the ever-increasing technical complexity of modern buildings and the client's requirement for speedy commencement and completion, which has led to a demand for more sophisticated and systematic methods of selection to be devised.

Franks (1990) rates each of the procurement systems in terms of their ability to satisfy the seven basic performance requirements, or expectations identified as being common to the majority of clients. Table 12.3 replicates this method of rating four systems. The ratings relate to a scale of 1–5, where 1 is the minimum and 5 is the maximum in terms of the individual system's ability to satisfy the listed requirements, and they are, of course, subjective.

This method of evaluating procurement systems and contract strategies by scoring how each system is able to meet the client's needs and project objectives is now reasonably well accepted in the industry, with these scores then being individually weighted according to their relative importance to an individual client/project. The total of the weighted scores reflects the suitability of the various systems for the project under consideration.

The example shown in Table 12.3 illustrates the point that this process of evaluation can only provide guidance in selecting the most appropriate system; each option will have some disadvantage or element of risk associated with it, but some will be better suited than others.

While the design and build option appears to be the most suitable in this example, this is unable to maintain a direct link between the client and the design team, although this weakness may well be overcome by using a variant of either of the two basic systems that do exhibit this characteristic.

As in the majority of cases, when using this method of assessment, there is unlikely to be a clear-cut decision and managerial judgement will need to be exercised in order to determine the final choice. In this context, other factors – particularly the ever-changing STEEP factors (social, technical, economic, environmental and political) will also affect the decision-making process.

Such methods of selection that have so far been described are undoubtedly useful in establishing those procurement systems which would *not* be appropriate, and the actual choice of system is therefore often reached by a process of elimination. They are, however, based on a very limited set of criteria. In addition, the procurement systems that are used are not reflective of the very wide, and sophisticated, range of procurement options that are currently available.

However, it should also be appreciated that a contingency approach to the procurement process is also relevant. In other words, there are likely to be, on the same project, many different suitable procurement routes, any, or all, of which will lead to a successful outcome provided that all other aspects of the project strategy have been dealt with correctly and that the project team performs to an appropriate standard.

There is no 'best buy' among procurement systems. Client organisations are complex, and different categories of client require discrete solutions to their procurement needs. Whereas on a well-planned and -managed project there may well be more than one

Table 12.3 Weighted selection model for choosing an appropriate procurement system

Procurement system selection

Standard criteria	Project requirements	Weighting	*Project manager's contract scoring*			
			Traditional	Design and build	Management contracting	Construction management
Timing	The project must be completed within 30 months. The decision to proceed will be made within 3 months.	4	4	10	8	8
Variations	The brief is well defined. It is unlikely that there will be major changes after construction commences.	1	8	4	8	9
Project nature	The building design is similar to those constructed for other departments in recent years.	2	8	5	8	8
Quality	Specification for the building is high but not prestigious.	2	8	5	8	8
Price certainty	Total costs must NOT exceed client's budget.	4	8	10	2	2
Professional responsibility	Project sponsor requires direct contact with design team.	3	10	1	10	10
Risk avoidance	Project sponsor wishes to pass controllable risks to other parties.	3	5	10	5	4
Responsibility	Minimum contractual links preferred.	2	5	10	4	1
TOTAL			143	157	133	133

system that will provide a satisfactory outcome, the choice of system(s) should always be made by matching the client's characteristics, objectives and project criteria to the characteristics of the most appropriate procurement method(s).

12.3.2 The selection of building procurement systems in practice

As we have mentioned previously, the choice of procurement system is a crucial strategic decision to both the client and the project, equal only to the establishment of the client's objectives and the decisions taken on the nature of the end product.

An examination of the procurement system selection process shows that it meets most of the criteria necessary for it to be considered as non-programmable or unstructured. It is complex, extends over a considerable period of time if carried out correctly, is often constrained by lack of information, is likely to be influenced by a number of individuals and the outcome is often unpredictable.

In addition, the process meets a number of the criteria which differentiate it from lesser decisions, e.g. it is comparatively novel, commits substantial resources, triggers other lesser decisions, and can be organisation-wide in its consequences. The process also reflects the characteristics of practical decision-making and choice in general, in that much of it takes place in an environment within which the constraints and consequences of possible actions are not precisely known.

In the past, the way in which clients chose their procurement systems was generally unconsidered, automatic and lacking in rigorous discipline, mirroring the disorderly nature of actual decision-making previously identified. Clients of the construction industry, having taken the primary decision that there is a need to build, would be faced with a choice as to which method of building procurement was the most appropriate in the circumstances and often merely accepted the advice of their consultants. Let us now look at this decision process in more detail.

Timing of the selection process

While the best advice from industry reports stresses the need to select the procurement route before appointing any individual or organisation, other than a principal adviser, many clients do not appear to recognise the necessity of making such an early decision or even realising that such a choice is required.

Over 75 per cent of experienced clients do choose the procurement system within the inception or feasibility stages of projects; i.e. sufficiently early on in the procurement process to enable any of the procurement systems to be chosen and to ensure that the client has no need to appoint any individual or organisation, other than perhaps an independent adviser, often prone to giving subjective advice on the choice. However, this means that a quarter of clients generally make their choice too far into the process – some of them much too far – by leaving the decision until the detailed design stage is under way and any possibility of an unbiased choice has been removed by the appointment of design consultants.

While a substantial majority therefore appear to be following recommended practice, the number of experienced organisations that do not make their choice early enough is still too high for good practice, particularly bearing in mind that these are experienced clients.

Setting objectives

There is little doubt that the majority of decisions to build, by their very nature, are made as the result of deliberate managerial strategies formulated within the aims and objectives of the organisation. Establishing whether this is the case with the decisions made about the selection of procurement systems is more difficult. The reports of various working parties established by the UK government to inquire into the problems of large industrial construction sites found that clients' policies/aims in relation to the implementation of their projects, including the selection of procurement systems, had not in general been designed to meet the unique circumstances of large, expensive, complex and lengthy projects and recommended that 'balanced and compatible policies' needed to be formulated for such projects.

Most clients are in fact aware of their role in defining the scope, objectives and priorities of a project, but the decision-makers within those organisations are often too concerned to maintain the organisation's policies and standing orders, which may dictate procedures such as compulsory competitive tendering, or will take the less-risk option and go for the route which has been used before, even if it not entirely suitable for the project in question. Reasons for this are given as 'official policy' or 'normal practice'. Many clients are also unaware that they have an essential contribution to make to the success of their project by developing a clear understanding of their project goals as early in the process as possible.

There is a natural tendency in all walks of life to use well-tried and trusted methods, and procurement is no different, especially if the 'normal' method has been developed to suit the client's own corporate objectives and resources. This point is probably more important in relation to experienced and semi-experienced clients, because they will tend to be larger and therefore more bureaucratic, whereas smaller companies may be experiencing the building process for the first time and their decision-making procedures will be more flexible and centred on a smaller number of people.

While these findings are not unexpected, particularly as inexperienced clients will not have had any previous opportunity to establish policies on procurement, they do highlight the fact that many experienced and partially experienced clients will be restrained by the policies of their organisation when making their choice of procurement system, although not normally to the extent of stipulating the use of a specific system or systems.

Examining the setting of objectives for the delivery of the project itself, whether affected by the client's organisational policies or not, again presents some difficulty because of a lack of the availability of definitive information on the practical implementation of this element of the decision process.

Many clients experience difficulties in determining what the 'problem' actually is and/or setting objectives to deal with it within the project brief. This is why it is essential that the lead consultant or project sponsor is skilled in constructing the brief, to tease from the client exactly what the project should satisfy. It is often the case that the briefing process in construction projects is inadequate, with the necessary information either not being forthcoming from the client or being assumed by the design consultants, resulting in an inefficient construction process from the very beginning.

The client, though, is not a construction expert, so it is not unusual for them to be unskilled in developing project briefs. They are manufacturing companies, retail organisations or service companies and have often voiced their dissatisfaction about the help given by the industry in developing their project brief, particularly bearing in mind the client's lack of knowledge of the industry and their lack of resources to carry out the formulation of the brief, which can often be just a very basic statement of need followed by an informal process of consultation and negotiation. This statement of need is often technically focused, concentrating on the 'hard' or physical outputs of the project and paying little or no attention to the softer organisational issues and not engaging in sufficient dialogue with their consultants.

The setting of project objectives by the client has been examined by a number of reports over recent years. For example, the working party that prepared the Wood Report in the 1970s found that the criteria necessary for a successful project that were consistently mentioned by all participants in the case study were: meeting the budget cost, low maintenance costs, time, quality, functionality and aesthetics (National Economic Development Office 1975).

Finally, it is always necessary to be aware that the temporary multi-disciplined organisations that manage projects in the construction industry are characterised by the fact that the client and the other participants in the process have to resolve disparities between the two levels of objectives that are always present in such circumstances, i.e. the specific and short-term project goals and the general and long-term objectives of the organisation to which the participants owe allegiance.

Identifying and developing alternative procurement system solutions

Without going into too much detail on the theory of decision-making, all decisions in organisations generally follow one of the following patterns:

- *Analytical search* – where the client's needs were analysed and project criteria were established and matched to the characteristics of the most appropriate procurement system. Included in this category were those decision processes where clients also took advice from external consultants, in-house experts and colleagues. This process follows the recommended decision-making route described previously.
- *Consultative search* – no formal analysis of needs or determination of project criteria were made, but a search of available procurement methods is conducted by a process of discussion with advisers and other experienced client organisations. A number of variants of this procedure can be identified which relate to the sources of advice; from the organisation's own managers (some of whom may have past experience of implementing construction projects), external construction-related consultants, other friendly organisations and building contractors. It is always possible that the sources of advice may have themselves carried out some form of analytical search in order to provide the guidance.
- *Historical evaluation* – the procedure whereby past experience of the organisation in implementing building projects, or the knowledge of an individual manager of such activities from previous employment, is used to evaluate the alternative solutions and develop the most appropriate solution.
- *Intuitive evaluation* – in this procedure, there is little search for alternatives or external advice and the identification of alternatives and the development of the eventual solution will be based solely on an intuitive evaluation of the problem of selection.
- *Policy compliance* – this procedure is used where a mature company policy is

established, often as a consequence of financial regulations, to control or restrict the use of procurement systems to those that satisfy the policy, for example, compulsory competitive tendering.

Research conducted in the past shows how the five classifications of the evaluation processes were distributed among the various categories of client. The recommended theoretical method, that of *analytical search*, was carried out by about a quarter of clients, with *consultative search* and *historical evaluation* by about one third each and the remaining methods accounting for the rest.

Regarding the final choice of route, at the end of this decision-making process, together with commitment to a specific alternative, there are fundamentally three different types of choice strategies:

- *Fulfilment choice* – where the needs of the client and/or the project criteria were seen to be fulfilled by the chosen procurement system. This routine includes complying with company policy, satisfying the need for single-point responsibility and the desire to use the organisation's own in-house designers.
- *Advisory choice* – where the choice is made on the advice of external consultants and internal experts or experienced employees, contractors and other friendly experienced organisations.
- *Historical choice* – where the choice of procurement method is made mainly, or solely, on the basis of previous satisfactory experience of a specific system. The use of this routine can be coupled with taking advice from external and internal 'consultants', reinforcing the view that there may well be some overlap between this method and the advisory choice category.

Generally speaking, it is the private/partially experienced or secondary clients who use the historical choice route to make the majority of their decisions, whereas private/inexperienced clients tend to favour the use of the advisory choice routine. This reinforces the common-sense conclusion that inexperienced clients make decisions on the basis of expert advice, or even by delegating the decision itself to such experts, while experienced and partially experienced clients tend to base their choices on personal experience. This is mainly because construction is a complex, technical and uncertain industry and if clients are not familiar with the technical aspects or cannot deal effectively with the uncertainty, they will naturally leave decisions to their consultants, which is what they are paid for!

Often, when clients use the historical choice routine, uncertainty and risk are ignored, presumably on the grounds that previous experience meant that such difficulties could be avoided as they had been in the past. This approach gives more credibility to such experience than is perhaps warranted but is a common corporate technique. Uncertainty is, however, present at both the evaluation and choice stage as a consequence of a lack of information about future events. Clients often rate the elimination and/or reduction of risk as being of importance when selecting procurement systems but still fail to consider it seriously, especially when getting the project started is of crucial importance.

12.4 Successful building procurement route selection

The success of a project can only be determined by the degree to which it meets the client's and/or the end-user's needs and requirements. With a very competent and experienced project team that has worked together on many similar projects, the choice of procurement system will not always be critical to achieving this client satisfaction. Therefore, it is necessary to select the procurement system with extreme care by following the procedure described in previous chapters:

- Identifying the characteristics and culture of the client and that of their organisation, establishing their needs and requirements, together with those of the end-user of the facility, any internal and external constraints and risks associated with the proposed development and, of course, the project objectives.
- Ensuring that all of the currently available procurement systems together with their respective characteristics, advantages, disadvantages and means of implementation are identified and known to the person or persons who will be responsible for the selection of the procurement system.
- Choosing the appropriate system by matching the client and project profiles to the method of procurement with the most suitable and compatible characteristics.
- Monitoring the effectiveness of the selected procurement method by obtaining feedback from all members of the project team.

12.4.1 Clients' characteristics, needs and project objectives

Clients' characteristics and culture

The culture of any client establishment is determined by its history and ownership, size, the technology it uses, its goals, objectives, environment and people. For example, the informal, relaxed, casual-dress management style of the typical, young, founder-dominated, US-based software house and the more formal, conservative ethos of a long-established, major high street bank is likely to result in very different approaches to the way in which they each handle any major project despite the fact that they can both be categorised as private/experienced/secondary clients.

It therefore follows that although the characteristics of the client and organisation will affect the way in which project-implementation categorisation is approached, i.e. into which of the six previously defined categories the client falls, they can only be used as an initial guide to the likely characteristics of any client.

Any externally, or even internally, appointed project manager/project sponsor will need to be aware that organisations are complex bodies and that, after placing the establishment into a particular category, it is also necessary to determine the specific characteristics, culture, policies and philosophies in order to refine the client's profile.

The complexity of many organisations, which may not necessarily be related to size, also makes it essential that the culture and characteristics of the major departments, and their management, are identified, particularly if they are to be stakeholders in the proposed project. In the same way, details of outside interests – such as consortia partners, financial funders, major shareholders and insurers, all of whom may well have a direct or indirect influence over the organisation's approach to project implementation – should be identified and included in the profile. This is effectively using the principles of supply-chain management to address the demand side of the project – maybe the term 'demand-chain management' is too obtuse.

It is easy to lose sight of the fact that in the case of many construction projects the client/owner usually sees only the end product and often has very little involvement with, or knowledge of, the actual project-implementation process. This is particularly true when the client is inexperienced or only partially experienced and therefore relies on their professional consultants. The principal adviser should then ensure that the client is made aware of the principles of the project-implementation process, the difficulties and dangers inherent in it and the various ways in which the successful procurement of the project can be achieved. The client's reaction to this information should provide an interesting insight into the culture of the organisation and the characteristics of its management, and should, hopefully, ensure that the client is less likely to be surprised when faced with the reality of the unique way in which construction projects are implemented when compared with the methods used to produce and provide products and services in the industrial and commercial sectors of industry.

Clients' needs and requirements

As we have stated several times, the client's requirements need to be established in advance of the project objectives and will not only embrace the spatial, functional and quality needs but also, most importantly from the point of view of the procurement system selection, the way that the client wishes the project to be structured, particularly in terms of risk, any internally or externally imposed constraints and any requirements stemming from the influence of external stakeholders.

Reference has already been made to the need to identify the 'real' client when attempting to establish their needs and requirements and the difficulties that this can cause when the project is to be used to satisfy the needs of disparate and possibly conflicting sections of the client organisation. On a project where the end-user is different from the owner, for example where the latter is a developer and the former a tenant, this situation is exacerbated and will make the determination of the needs of the eventual occupier a difficult and sometimes near-impossible task.

In terms of corporate decision-making, there are seven different roles, first developed by Cyert and March in the early 1960s:

1 *Policy-makers* – usually senior members of the organisation with authority and responsibility for policy of what to purchase and how, e.g. the directors of the business.
2 *Purchasers* – these could either be the administrators filling in the purchase orders or requisition forms, or those responsible for the final recommendation.
3 *Users* – those who ultimately use the product/service. May not be part of the purchasing organisation, e.g. tenants.
4 *Appraisers* – those with specialist knowledge who can appraise the service and advise on appropriate key performance indicators (KPIs) of the facility. Appraisers may even have defined the KPIs.
5 *Influencers* – those who can influence the decision-making process by providing information and criteria for evaluation. These may be internal or external to the organisation.

5   *Gatekeepers* – those who control the flow of information to others within the firm and decision-making unit. Never underestimate the power of secretaries or PAs.
6   *Deciders* – those who finally make the decision at the end of the process.

Cyert and March 1963

Despite these complications, every effort needs to be made to establish what the eventual user of the project requires, although on occasion this can conflict with the objectives of the owner, and diplomacy and tact will be required in order to establish priorities and/or a compromise between the different parties' requirements. It is therefore essential to ensure that the interests of the various internal factions are accommodated and that the final project brief reflects all of their requirements, or any compromises reached, and is 'signed off' by all of the interested parties.

Some clients' requirements will in themselves be constraints upon the way the project and especially the selection of the most appropriate procurement system can be carried out. As we have seen, the internal environment of any category of client organisation can impose constraints upon the project. This is most likely to be in the form of restrictions stemming from the policies or standing orders/internal regulations, or even unwritten rules, of the organisation.

While the most obvious examples of restrictions, such as the use of certain specific tendering strategies, types and forms of contract, have already been discussed, other restraints can range from restrictions on the use of certain consultants, contractors and procurement systems to the imposition of unrealistically low financial approval limits by public-sector authorities on clients' project managers.

All such constraints are likely to be to some degree detrimental to the successful and efficient implementation of the project, but many organisations, even those that appear to be extremely commercial in all their other dealings, are often not willing or able to waive any internal regulations. Some clients, possibly those in the private rather than the public sector, may be persuaded by an exercise which compares the cost of adhering to the policies with the lower cost that would be incurred if the restrictions were removed or amended.

In many cases, reluctance to amend or waive internal policies or regulations stems from the organisation being unprepared to accept any real, or perceived, risk that any divergence would create. Risk analysis techniques can be used to assess accurately the level of risk, thereby enabling financial/time comparisons to be made which might serve to persuade the client to adopt a less conservative approach to amending or waiving the restraint.

Project funding and funders are one of the more common areas for constraints on the project organisation, particularly since the 'credit crunch' of 2008–10 has altered the way that financial institutions view risk, and this is even more critical if the funder has its own particular expenditure programme requirements or requires specific spatial, functional or quality standards to ensure market acceptability should disaster strike and the project reverts to the funder's ownership.

Clients who are part of central or local government are well known for being restricted by various policies, regulations, standing orders, etc., which are there to ensure public accountability, but they are just as often constrained by funding being restricted to a specific time frame, their inability to roll over funding from one year to the next thus creating excessive workload at the end of the financial year, the drip-feeding of funding for long-term projects and the general inflexibility of central and local government funding systems.

As most funding-related constraints cannot be removed without great difficulty, it is usually only possible to mitigate the consequences by selecting methods of procurement which do not inhibit expenditure from being incurred at times when money is available or even for other members of the project team to provide short- or long-term finance, as with private finance initiative (PFI) schemes – see Chapter 11. However, such devices do prove costly and may even detract from, reduce or possibly invalidate the viability of the project. As a result, these devices must be determined as part of the project strategy and incorporated into any requests for tenders or proposals.

Some physical constraints can also affect the choice of procurement system. For example, the presence of a number of existing occupied buildings on the site which are only going to be released in stages and thus become available for demolition or refurbishment, could well determine the way in which the project is procured and which form of contract is most appropriate.

As mentioned previously, the question of risk in construction projects also needs to be considered at a very early stage in the project's life in order that these risks can be identified and responsibilities for accepting such risks determined. These decisions will in turn determine which procurement systems are compatible with the level of risk that the client wishes to take upon themselves or pass on to others. Figure 2.2 in Chapter 2 shows how the different procurement routes can be graded depending on the financial risk allocated to the client or the contractor. This ranges from virtually 100 per cent acceptance of risk by the contractor when using the turnkey or package deal system to the same rate of acceptance by the client when using the pure form of construction management. Clients who are therefore risk-averse are considerably reducing the number of procurement systems that are available to them, whereas organisations that are prepared to accept risks that they are able to manage effectively themselves will consequently have a wider range of systems to choose from.

Project objectives

During the project inception stage, when the enthusiasm to get going is at its height, it is often forgotten that the object of the exercise is to satisfy the client by ensuring that their project objectives are met. As in most other human activities, this can be best achieved by spending sufficient time planning and preparing for the implementation process itself; the formulation of a considered and comprehensive project strategy will ensure that this is done and that success is finally achieved. Time spent on planning is never wasted.

The project objectives will not only be determined by the specific requirements and aims of the project itself but will also reflect the organisation's wider strategy and long-term business/social/political aims, i.e. the client's requirements.

Difficulties often arise during the project definition stage, when narrow project objectives conflict with long-term corporate goals. Departmental, divisional or subsidiary companies within large organisations or groups may have their own particular agendas and accidentally or deliberately overlook, or simply not be aware of, the wider corporate/political or social issues. All conflict, whether it is among departments, divisions or individuals within an organisation creates problems and adds to project costs and time. Once again, the need to identify the real client is of major importance when establishing the project objectives and satisfying the demands of disparate entities within an organisation.

The control of the design stage and the construction stage needs to be carried out against these sets of aims and objectives, which will be set in stone in the project brief document. This document, as well as recording the client's functional, spatial and technical needs, should include details of the required timing, cost parameters and quality/functionality standards. As we have also mentioned earlier, an accurate, definitive and timely project brief will enable an early decision to be made with regard to the choice of the most appropriate procurement system. Too often, an early decision is made on building projects to appoint a team of design consultants rather than a single unbiased principal adviser, with the result that a procurement system is often chosen by default rather than design.

The three primary objectives of time, cost and quality/functionality need to be closely defined in the project brief to ensure that the specific project objectives are accurately identified.

Some clients, and principal advisers, may be concerned at divulging details of cost estimates, but little is lost and a great deal gained from all members of the team being aware of the financial parameters that have been set, and the majority of competent and professional team members will objectively endeavour to better rather than just meet these targets. To be told that the client requires minimum cost is insufficient. An actual maximum cost budget could easily be set, a time schedule for expenditure specified, a contingency sum agreed, accountability requirements defined, etc., as all of these requirements could impinge on all, or some, of the other elements within the project strategy. In the same way, expressing time in simplistic terms, i.e. 'the shortest time possible', is equally unacceptable. Project duration, the design period, the required commencement date, sectional handover, the length of time allowed for construction, the length of the defects liability period, etc. all need to be specified when they are appropriate to the client's project objectives.

Quality and functionality should be expressed in specific terms using appropriate accepted industry standards in written specifications, sample materials and panels, room mock-ups, etc. The achievement of functionality will of course need to rely eventually on drawings and other visual communication, but in the early stages of a project, performance specifications rather than prescriptive specifications could be used to establish the client's requirements.

All of the three primary objectives need to be prioritised and weighted. This is a task which many clients, particularly those with little experience of project implementation, find difficult but which is vital to the eventual success of any project. The principal adviser should therefore be skilled in teasing out of the client their priorities and critical requirements.

Once the primary objectives have been prioritised and weighted, a number of procurement systems will have been found to be inappropriate for use on the project under consideration. This number may be further increased when the project's secondary objectives have been considered.

Although they are of lesser importance than the three main objectives, in combination the secondary objectives can have considerable influence on the project procurement process and project success. Many clients have, for example, a preference for carrying out the installation of their process or manufacturing equipment during the currency of the main construction contract. Whereas in practice this is perfectly possible, problems can arise if the client's requirements in this respect have not been fully considered at the beginning (such as the location of builder's work in connection with the client-supplied and -installed components) and comprehensively incorporated within the tender documentation and subsequent contractual arrangements. If there are other contractors on site over whom the main contractor has no responsibility and who are not carrying out part of the main contract works, there are issues regarding liaison, health and safety, welfare and insurance to consider, to name only a few!

If the client wishes to operate the facility with a trained staff immediately it has been handed over by the contractor; wishes to deal with only one single organisation during the project duration; wishes to use specific design professionals during the early stages of the project and then place the responsibility for detailed design and construction onto a contractor, etc., these apparently secondary requirements will have a significant effect on the category of procurement system that they will be able to use.

Should two or more of these examples of secondary objectives be required in combination on a single project, they would be likely to have such a major effect on the procurement process that they should be redesignated as primary rather than secondary. The client's secondary needs should also be prioritised and weighted.

The proper and early identification of the client's objectives is a vital step in formulating the project strategy and selecting the most appropriate procurement system, thus ensuring the effective control of the procurement process and the management of the project as a whole. Time spent on ensuring that objectives are properly and accurately identified at this stage will be repaid many times over during the implementation of the project.

12.4.2 Identifying all currently available procurement routes

Very few clients, even those with experience of building, are aware of all the procurement systems or the way the systems themselves operate. Indeed, the procurement routes together with the choice of contracts for that route have progressed so quickly in recent years that even the average consultant may not necessarily be fully aware of the full range of potential methods or their characteristics.

We also know that many clients appoint their design consultants and possibly contractors during the initial stages of the project, when such pre-emptive appointments can only hinder the unbiased selection of the most appropriate procurement system. These appointments should take place only once the most appropriate procurement route has been chosen. It is therefore suggested that clients should look to the, hopefully unbiased and independent, person whom they have appointed to advise on, or carry out, the initial management or co-ordination of the project to identify all of the suitable procurement systems.

This exercise will entail not only identifying all of the appropriate methods but also obtaining details of the way the individual methods are implemented, the characteristics of the various systems and their advantages and disadvantages. This information is now quite readily available from published sources, the internet, government departments, many of the construction industry's representative bodies and professional institutions.

Ideally, a short report, or discussion paper, containing all of this information should be prepared for, submitted to and discussed with the client to make them aware of the methods that will be considered at the next stage in the process and to obtain their opinions on the best way forward.

Once the final list of potential procurement systems has been agreed, it will be time to move on to the most critical stage of all – selection of the most appropriate method.

12.4.3 Choosing the most appropriate route

Clients often do not select the procurement method they eventually use in a disciplined and logical way. If success is to be achieved with minimum difficulty, the decision about the most effective means of procuring the project must be made by comparing the client's characteristics, needs and project objectives with the characteristics, advantages and disadvantages of each of the selected systems. Whichever of the systems is most compatible is then chosen.

Despite the documented reluctance of clients to use the aids to selection that have previously been described, these techniques can be of great assistance in reducing the number of potential procurement systems to more manageable levels and, in some cases, in selecting specific methods as being most suitable for the project in question.

Once the choice has been made, the client will need to be advised not only of the decision but most importantly of the contractual, practical and financial consequences of using the selected method, in order that formal approval may be obtained before any further action is taken.

It has always proved difficult to forecast which procurement route would produce a cheaper or more cost-effective building. Because each project is unique, in terms of its location, size, shape and timescale, it is all but impossible to compare the different procurement routes taken on different buildings. Various studies have purported to show that design and build can be up to 20 per cent cheaper than traditional procurement, but care must be taken in relying on such figures.

Once the final selection of the most appropriate procurement system has been made, the client needs to consider the nature of the environment in which they wish to implement the chosen method. Do they wish the project to be carried out within a co-operative model or on a purely contractual basis?

Even under the comparatively ideal circumstances encountered in manufacturing industry and commerce, persuading individuals or disparate groups to work together in a spirit of partnership towards a common goal is never an easy task and demands the presence of supportive policies, above-average management skills and a project organisational structure that is both flexible and capable of dealing with cross-departmental operations. As we saw in Chapter 1, the construction industry is fragmented and made up of a lot of relatively small firms who are used to competitive tendering, confrontation and closed-book accounting. Construction projects therefore have been historically unable to meet these criteria, and to change the culture now requires a massive effort. Perhaps the effect of the post-2008 recession will give the necessary shock. Time will tell.

We have also seen that project teams in construction usually consist of individuals from different organisations, each with its own agenda and objectives and more often than not unknown to each other before the project starts. This would be sufficient by itself to account for the difficulties experienced in managing projects, but in addition to this constraint, the tendering arrangements, conventional procurement systems, adversarial types and forms of contract, the unique nature of construction projects, the difficult physical environment in which work is carried out and the combative attitude of many managers mean that the establishment of a positive, co-operative and supportive approach by members of the team is extremely difficult.

Clients have a very important part to play in overcoming these difficulties, initially by establishing an environment in which co-operation can flourish and in which positive, objective attitudes are encouraged. Even without the use of non-adversarial contractual arrangements, a positive environment can be created by any client organisation willing to adopt appropriate policies and attitudes and to ensure they are communicated, established and adopted throughout the project team, and in every aspect of the project's life, in order to create a constructive project culture.

This has always been achieved on a relatively small number of projects by the simple expedient of the client being constructively and continuously involved with the implementation process and by applying the same enlightened philosophies and approaches as those adopted within their own organisation coupled with the use of tried and trusted consultants, contractors, suppliers and subcontractors and utilising the most appropriate procurement systems and contractual arrangements.

Where such an approach has been used, the project has generally been successful, because the client has always been satisfied that their objectives are being met. All the other members of the team have benefited by achieving their short-term commercial goals and the long-term aim of satisfying an important client and heightening the chances of their services being used on future projects.

Unfortunately, such far-sighted and sophisticated clients, or project teams, are not common in the construction industry. In order to achieve the same level of success more generally, it is necessary to formalise this enlightened approach into some form of arrangement similar to that now known as 'partnering' and/or the use of non-confrontational procurement systems such as those in the management-orientated category (see Chapter 9).

These techniques have been shown to produce a 'can do' attitude among the participants and to foster and require a co-operative approach to problem-solving among the client and the team members, and, in conjunction with the use of the correctly chosen procurement method, help to ensure a successful project.

Communicating the choice of procurement system once it has been made to all of the parties that are participating in the project at this stage is often overlooked, but the inclusion of the decision within the circulated project brief should ensure that all of the participants are aware of the choice that has been made.

Depending on the nature of the client organisation, and such external members of the project team that have been appointed, it is often beneficial for the consequences of the choice, in terms of the advantages and disadvantages of its use and the constraints that it places on the behaviour and involvement of the current participants in the future implementation of the project, to be spelt out to all concerned and confirmed in writing as appropriate.

12.4.4 The importance of M&E building services

Mechanical and electrical building services (also known as MEP in international contracts – mechanical, electrical and plumbing) typically account for anything up to 40 per cent of the capital value of a large commercial scheme and the design and installation are often major components of the project's critical path. On some specialist projects, such as high-tech research laboratories, the value of the services component can exceed 75 per cent, and the requirements for total reliability and fail-safe systems can place great demands on the ability of the services team to deliver.

Irrespective of the scale and complexity of the systems, good-quality design and installation work, effective co-ordination and rigorous commissioning are essential for the long-term, efficient operation of the asset.

Building services require a high degree of co-ordination with other elements of the building, as the various pipes, ducts and cables need to be fixed to walls, ceilings and floors, holes need to be made to allow these services to pass through the walls and floors, and finishings need to be adapted around the switches, sockets, fuse boxes, etc.

Owing to this interrelated nature of the design and construction of building services, the decisions made by clients in determining the overall procurement strategy will have a substantial effect on project organisation as well as project outcomes.

The client therefore needs to consider a number of key issues, which include:

- Who will be responsible for design at the various stages of the project: the consulting engineer, a specialist services engineer, the principal contractor or the specialist services contractor?
- Who will co-ordinate the design and installation of the elements of the works?
- If there is a separate requirement for single-point responsibility for building services, how will this be reconciled with the main contract works?
- What degree of control over the selection of specialist contractors is required by the client? What will be the impact of this requirement on the project programme and tendering procedures?
- To what extent is price competition required by the client in the appointment of the services supply chain?

Although many of these issues are common to general procurement and have been discussed in previous chapters, the extra requirement for a substantive completion of building services design before construction starts combined with the requirement to co-ordinate the works of specialist trades is often incompatible with the objectives of the traditional procurement routes and especially design and build, which is based on sequential design transferred from the design team to the 'main' contractor.

Experienced and informed clients, who generally understand the overall contribution that services contractors are required to make, and who are able to create the conditions where designers and contractors can work together effectively, will usually get the best results from the services components of the project.

As already described, much of the difference between building services procurement and 'normal' construction is concerned with the allocation of design work and responsibility for the design, which is often carried by a separate design engineer or the specialist subcontractor. Furthermore, the fixed sequence of how design is developed, from the initial calculation of requirements, through the selection of equipment, the sizing and routing of ducts and cables and the co-ordination of systems, generally requires the input of contractors, as the design cannot be completed in isolation from the building design.

Therefore, given the supply chain's high level of design involvement, the effect of decisions by contractors on detailed design and split responsibilities for co-ordination, there is plenty of opportunity for ambiguity, error and confused contractual responsibilities, which, as we have already seen, are the perfect breeding ground for disputes, claims and additional costs to the client as well as late delivery.

A co-ordinated approach to the procurement of building services should seek to address this issue. In dealing with these problems, the client or their representative should ensure:

- Clarity in defining the scope of the various consultants' appointments. This requirement extends beyond a clear statement of the consulting engineer's duties, and should also include an understanding of which parties will undertake the various duties and how they are expected to coordinate and communicate together.
- Clear understanding regarding the extent of the design and co-ordination roles of the various parties, together with requirements for obtaining approvals from the client and any regulatory bodies. Often these duties are inferred, which can lead to misunderstanding and poor performance by both the consultant and contractors.
- Clear definitions of the contractors' design roles, which should also identify constraints that might affect the completion of design work on time. A detailed schedule of drawings, schedules of equipment, etc. which define the scope of work can be valuable in this respect, but needs to be made available early to meet the demands of the design and co-ordination programme.
- A clear programme for the design and construction sequence of the services installations, which must make provision for design activities required after the contractor's appointment, so that there is sufficient time for the completion of detailed design, design co-ordination and testing and commissioning activities. As many of the parties responsible for providing these services may not have been appointed when the initial programme is drawn up, the skill of the programmer in anticipating the sequence and duration of activities required is critical.

As a result, it can often be difficult for clients, particularly inexperienced clients, to be able to determine the scope of service they require at the outset of a project, or to understand the implications of the terms of engagement they may have signed up to. Some standard forms of engagement for consultants may not include all the services which are required on a particular project

As building services are generally carried out by specialist contractors who are usually in a subcontract relationship with the main contractor (even though they may represent up to 75 per cent of the value of the works), the specialist subcontractor faces all the same challenges of the main contractor, and in addition, the work is delivered in ways that can add further complexity and uncertainty. Some of the common problems experienced on projects are now briefly considered.

Drawing standards

Even if the consultant's and/or contractor's design duties and co-ordination responsibilities are well defined and agreed, there can be problems with the completeness and timeliness of drawn information by parties to the contract, if there is no fixed definition of the information required at a particular stage.

Physical constraints

As the design progresses to detailing layouts and suchlike, problems associated with the physical constraints of the building may occur. Although a co-ordinated project team will generally devise a solution that provides sufficient space for equipment and pipe/ductwork, constraints on space may limit the options for equipment selection.

Co-ordination

Problems associated with design co-ordination may include dimensional clashes, on-site rework, poor system interfaces and poor design solutions around physical constraints. To overcome this, innovations such as 3D modelling can often solve many of the dimensional issues and design co-ordination. Another area where co-ordination difficulties can arise relates to the involvement of specialist systems which are required to work together, i.e. the design work has knock-on effects on other systems. The co-ordination of the systems relies on the painstakingly detailed scheduling of all connections, control systems input/output sizing etc. The aim is to ensure that responsibility for co-ordination is clearly identified and effectively managed. The preparation of such a co-ordination matrix is invariably not included in standard duties, and the requirement therefore often has to be specifically introduced by the client.

Design management

As mentioned in Chapter 4, managing of the design process is critical to the smooth running of the project, and effective co-ordination depends largely on the quality of the overall design and construction programme, the contractor's own design resources and the services contractor's design management skills. These are vital, as the contractor must ensure that design work by any equipment manufacturers and specialist subcontractors is completed on time and co-ordinated in accordance with the overall programme for approvals, procurement and construction.

Management resource

A further challenge for some specialist contractors relates to the experience and competence of their management resource and the main contractor's ability to manage the completion of all aspects of the subcontracted works. This can be a problem when the specialist packages are on the critical path of the project. Many services specialists prefer to work on package-based contracts, as they give direct access to the designers and client. How the client manages these specialists will depend largely on the procurement strategy it adopts. Typically, the choice is between a lump-sum approach, with management and commercial risk held by a main contractor (traditional procurement – see Chapter 6, section 6.1), or a package route that has greater transparency but also more risk exposure on cost, scope and performance (overlapping roles – see Chapter 7). Where a package route is adopted, specialist contractors are an area of particular risk in terms of performance and co-ordination. The client may need to invest further in programming, design management and post-contract monitoring.

Effective contractor involvement

As mentioned in previous chapters, it is generally considered beneficial to involve contractors at the earliest opportunity in the procurement process, so that they can contribute to design and buildability reviews and the selection of appropriate plant and equipment. Some consultants consider that this early involvement only generates benefits to the client and project if the design work is sufficiently advanced to enable the contractor to develop the detailed design without the risk of abortive work. However, this may miss the point, as the designer is still responsible for the design, and the contractor would only give their practical advice, and the earlier that is given the better.

Programming of works

According to the Building Services Research and Information Association (BSRIA), UK building services contractors achieve comparatively low levels of efficiency once on site. Much of this is related to issues with the quality and experience of site labour, but also concerns the way in which site works are programmed and managed. On many projects, the contractor is required to start their work as early as possible, when they may not have clear access to work areas, the design is not fully developed, etc.

Commissioning

This is about making sure that the systems actually work when they have been installed. Provision for commissioning must be made in all system design and is supervised by the project consultants together with any licensing authorities, and therefore must be properly resourced and ringfenced in the programme. The installation contractor will be required to guarantee the workings of the installation for a period of time, so full and rigorous commissioning is vital.

Faced with the risk of possible uncontrolled change associated with the delegation of design work, and requirements for co-ordination between systems specialists, how should a client go about the procurement of M&E (MEP) building services?

As mentioned above, the procurement strategy must be determined by the client's priorities – their experience and willingness to accept risk exposure, their preferences for working with certain firms and their favoured forms of contract.

Irrespective of the approach chosen, many clients would wish to increase their involvement in subcontractor procurement so that they have more influence over the selection of subcontractors and the basis on which they are appointed. In the current marketplace, where the 'nomination' of subcontractors is now almost impossible, negotiation rather than commercial tendering is increasingly common, requiring greater post-contract discipline to ensure that the contractor does not recover low tender/negotiated costs through claims for delay, rework and disruption. In arriving at a decision, the client should also consider the following broader issues.

Cost certainty

As mentioned in Chapter 5, some form of pricing document from all appointed M&E contractors is essential for post-contract management and must be obtained before the subcontract is placed.

234 The selection of building procurement systems

Design responsibility

A full range of warranties that reflect the duties of each party must be obtained from all consultants, specialist contractors, etc.

Design of work packages

If a work-package strategy is decided (i.e. construction management or management contracting procurement – see Chapter 7), a balance has to be struck between the size of the subcontract packages. Large packages will offer a diversified risk and efficient use of a contractor's resource, whereas a larger number of small ones will give the client direct access to the resources that deliver the work. However, the use of a lot of smaller subcontracts may expose the client to companies with poorer management skills and also require co-ordination with many more companies and interfaces. In most cases, it is preferable to use a smaller number of larger subcontracts.

Programming of the works

The programming of the date of design completion and the sequence of construction is another key factor. In lump-sum contracts, programming is generally the main contractor's responsibility. The efficiency of the contractor's programme and method statement represents a significant risk to project delivery and will be the major reason for differences in price between tendering contractors. Where the transparency of programming, sequencing of packages and the performance of trades are important issues to the client, and if the risks associated with taking an active role are seen as acceptable, a package-based approach may be an appropriate response. Modern building services procurement exemplifies many of the trends associated with current thinking on project management, encouraging collaboration and the use of specialist skills. Unfortunately, procurement practice has not evolved sufficiently to avoid some of the complications that come from the sharing of design roles, the co-ordination of complex systems and the exploitation of ambiguous contract documentation. To get the best out of designers and supply chains, clients must understand the reasons for current systems and the implications for the project if they do not address sources of weakness. Clients who proceed with the project without this understanding risk poor value solutions and are unlikely to get full benefit from their specialists.

12.4.5 Monitoring and feedback

All commercial activities need to be monitored in order that progress can be measured against the specific predetermined objectives, and if necessary remedial action can be taken to correct under- or over-achievement.

The monitoring of the effectiveness of the chosen procurement system presents some difficulty because of the qualitative nature of the task. The majority of the primary and secondary project objectives are capable of being measured against quantitative targets, but whether the choice of procurement system has been correct cannot be determined other than by the measures of success applied to these objectives and the final outcome of the project-implementation process.

During the implementation of the project, it is of course possible to obtain feedback on the way the system is performing from individual members of the project team, although it will be appreciated that such information is subjective. However, regular monitoring of and feedback from the team members need to be carried out against a formal set of criteria, and their comments noted.

The client's impression of the way in which the system is performing is of primary interest and importance and must therefore be monitored and recorded in the same way as the opinions of the project team. At the completion of the project, the initial criteria for success reflected in the client's needs and the project objectives should be the benchmark against which the performance of the project team and the procurement system itself should be measured.

The feedback obtained from the ongoing exercises will provide valuable information on the way in which the system is performing and will enable the project team to establish whether its performance is satisfactory. Unfortunately, should this not be the case, it is rarely possible for the choice of procurement method to be changed during the currency of the construction phase of a project without considerable difficulty in making the necessary contractual and practical modifications. Under such circumstances, any necessary changes will need to be taken by addressing the design, construction methods or management elements of the project.

Notwithstanding such difficulties, the guidance obtained from feedback as to the management of the procurement, and other aspects of the project, together with all of the information gathered over its duration will be extremely valuable and, once completion has been achieved, should be incorporated into a brief report and circulated to all members of the project team.

12.5 Summary and tutorial questions

12.5.1 Summary

Decision-making is a process which starts with the objectives or needs of the purchaser and ends with the selection of a product or service which they hope will satisfy those objectives or needs. An analysis of whether it has done so will help to improve future decisions, if indeed there are to be any future decisions. Figure 12.1 gives this 'decision loop' in diagrammatic form. In terms of theoretical decision-making, there are three basic models. First, rational decision-making assumes that the purchaser knows exactly what they want and has perfect information from all the various alternatives. Unlikely. Second, behavioural decision-making looks at each alternative on its own merits (rather than comparatively) and against a set of predetermined criteria. Again, unlikely, as almost all human decisions have some form of 'gut feeling' associated with them. Third, the political bargaining model creates the optimum satisfactory solution (called 'satisficing' in some texts) depending on the various factions in the decision-making process.

With these theoretical models in mind, the chapter looked at the ways that procurement routes are selected in the real world and how to create a decision matrix with the client criteria and appropriate weighting against the various procurement alternatives. Clearly, the actual route chosen will depend mostly on the client's priorities, as well as the skills and expertise of the tendering contractors under consideration. For experienced and regular clients, it is additionally crucial to monitor the success of the chosen route, to ensure improvement in the future.

12.5.2 Tutorial questions

1   If you were building your own house, what would be your priorities? Rank them in order of importance.
2   If you were building a house for somebody else (that you didn't know), would this order of priorities be different?
3   Answer questions 1 and 2 again but come to a collective decision with at least four other colleagues.
4   What factors would affect the procurement route decision in public-sector organisations and private-sector organisations?
5   What factors would affect the procurement route decision in experienced clients and inexperienced clients?

13 Future trends

13.1 The longer-term effects of the economic downturn on the construction industry

The crisis in the world's financial system, which started in 2008, has had a major effect on the construction industry and it now faces the toughest challenges that anybody can remember. Jobs are being cut, salaries are becoming more 'realistic' and the modern concepts of sustainability and partnering are being set aside as the core issue of survival concentrates the minds of the industry's leaders. Indeed, at the time of completing this book in late-2010 there is a very real risk of a 'double-dip' recession following further financial crisis, which may make credit even harder to obtain and force severe austerity measures on even the most developed economies in the world.

13.1.1 *The business environment*

The initial principal cause of the recession was a crisis in the banking industry caused largely by injudicious lending (called 'sub-prime mortgages' or 'toxic debts'). The banks and primary lenders packaging up and selling on these debts to other financial institutions in the 'derivatives' market further exacerbated this problem. Financial derivatives are not tangible products or services but are 'derived' from real products and services. Derivatives therefore produce a type of 'false market' where a growing toxic debt is traded in the financial market. This is until particular institutions are poisoned by the toxicity and collapse, then the natural cycle of economic growth begins to fall. The non-derivative or 'real' product was a long-term debt or mortgage that was never going to be paid back, and as a result it was not difficult to see that the bubble was going to burst at some point – the unknown factor was when this would happen. That point was reached when Lehman Brothers (a US bank) went into bankruptcy in 2008 setting off the whole train of events across the globe.

Banks were very nervous about lending money, even though that is their core business and they should have sophisticated risk-assessment strategies for client lending. Following the US difficulties, banks in the UK and Europe found that they also had 'sub-prime' lending and reacted by clamping down on credit and loans to even the most creditworthy customers. Mortgages became very difficult to obtain, with the obvious effect on house builders and construction firms in the sector. Banks also clamped down on builders as well as homebuyers, with bank credit now a scarce resource. The certainties that previously existed in the business environment disappeared almost overnight as the rules of the game changed and firms which did not have high cash deposits in their current accounts (known as liquidity ratios) found that they were in serious trouble. Banks appealed to the government for help and many were subject to takeovers by their larger competitors (as in the case of HBOS being bought by Lloyds TSB). Some banks were even bought out by the UK government (e.g. the Royal Bank of Scotland Group) and are now majority-owned by the public in what has become the largest nationalisation of the 'means of exchange' in generations.

Combined with a worsening credit environment, bankers were even more judicious when lending cash. Most construction companies with high borrowing were locked in serious conversations with their banks, which were now looking to put them on a tight leash. Those firms with high liquidity ratios clearly did not suffer as greatly and the situation brought acquisition opportunities for cash-rich contractors. In the housebuilding sector, many firms engage in the buying of potential development land years before they expect to build on it (called land banks) and were forced to sell some of this land to raise cash and keep going in the short term. It is unlikely that the credit crunch will trigger a wave of consolidation in housebuilding, as mergers/takeovers are normally to enhance earnings, which may only kick in a few years down the line, or are made by appropriating other firms' cash deposits.

As well as credit and liquidity, the business environment works on the basis of confidence. Clients need to be confident that the builder will do the work, and the builder needs to be confident that the client will pay the due certificates. Furthermore, the banks need to be confident that monies lent out will be paid back, or at least the interest due every month will be paid back. Many firms rely on this confidence in managing their cash flows. Nobody wants too much money sitting in their current accounts as these funds are not working for them; any spare funds should either be used to buy business assets or invested to make a profit. Firms should therefore keep only a minimum amount in their current account to pay for day-to-day costs until the next client payment comes in. When confidence evaporates, this causes major problems to the day-to-day running of the business and within the wider macro-economy.

13.1.2 The macro-economic picture

One factor underlying the construction industry's vulnerability to the downturn was that the public sector does not now have the power it once had to act as a Keynesian counterweight to the business cycle of boom and bust. According to the Construction Products Association (CPA), the public-sector market in 1970 was worth £27.5 billion and the private sector £28 billion, therefore making them about equal in terms of construction workload. In 2007, the public sector's demand had shrunk to £21.5 billion, about a third of the private sector's £61 billion. The reason for this was that much of construction work which had traditionally been seen as public-sector, such as social housing, was now undertaken (if at all) by housing associations, which are classed as private-sector, although, as we saw in Chapter 2, they are in effect the third, not-for-profit, sector.

Since 2008, the commercial property sector has had a steep fall and is unlikely to recover for several more years. This means that the construction industry has taken a big hit and will continue to suffer. However, leading the way in the public sector is education, and the Building Schools for the Future (BSF) programme, some have argued, has helped to prop up the industry. Some £22 billion was scheduled for spending on this programme between 2008 and 2011. Rather ironically given the programme's title, BSF's future is not entirely certain and in many cases come to an end under the new coalition government and their public-sector cuts. Health spending has also peaked and there are no large hospital schemes in the pipeline. Even more pessimistically, the new coalition government in the UK is likely to target public-sector construction as central to their austerity programme from 2010, so no public-sector-driven construction work is necessarily safe.

13.1.3 Small businesses

The construction industry is characterised by small businesses and fragmentation, as we saw in Chapter 1. During difficult times, they are the ones who suffer most, as their average projects or work packages are small, with short timescales and relatively low cost. On the whole, small businesses in the building industry do not have the larger projects to keep them afloat during difficult times. As previously discussed, this lack of business for small builders is compounded by credit conditions caused by the banking crisis. Small builders struggle, as they cannot get loans very easily from banks and in turn this puts pressure on their cash flow. Cash-flow problems are further compounded on small businesses when larger contracting firms do not pay them on time and only release payment sporadically. Meanwhile, the clients of small builders also find it difficult to borrow money, meaning that many proposed projects stay on the drawing board or current projects are put on hold.

13.1.4 Procurement

We have already seen in previous chapters that the credit crunch has had the effect of reversing the industry's trend towards negotiated contracts and partnering. Instead, there has been a swing back towards single-stage, lowest-bid tenders aimed at achieving the lowest possible out-turn cost. Tender prices barely rose in the period 2008–10, but material costs did rise significantly, which makes matters worse for hard-pressed contractors.

Changes in the labour market have also changed the landscape in the procurement of a building project's workforce. During the last downturn, labour costs were high, which gave contractors scope to save money by pushing them down again. Wage deflation (in the UK) is not happening now, in part as a result of workers from A8 (Accession 8) countries keeping wages low since the EU working-permit zone was further expanded. Wages have stayed low even though most of these workers have returned to their home countries. Although contractors are being less choosy about what work they take and the levels of risk they accept, clients are generally unsympathetic. They feel contractors have been getting a bigger share of their profits in recent years with the negotiated-style contracts. Now clients are more interested in going down the single-stage tender route.

13.1.5 Other changes

Greater efficiency of design

As all students of quantity surveying know, the shape of a building affects the wall–floor ratio. This means that for a particular floor area, the amount (and therefore cost) of wall construction will vary, with the square box being the most economical. As every student of architecture knows, don't listen to quantity surveyors when designing buildings or you will never get noticed as an architect. However, nowadays clients are looking for greater value in their buildings, and increasing the lettable floor area while reducing capital costs is a good way of increasing value.

So, will this mean a return to cheap, boxy buildings like those built in the 1960s and 1970s? Hopefully not, but the effect of a tax on unoccupied buildings and the development tax brought in by the coalition government budget in June 2010 will also have an effect on capital values. Additionally, the environmental agenda embodied in the revised Part L of the UK Building Regulations will affect architectural form as the move away from air conditioning and artificial light to natural ventilation and daylight will discourage box design with areas far away from an external envelope.

Build-to-let instead of buy-to-let

Because of the difficulty in obtaining mortgages, the buy-to-let market has been very deeply affected and few people have the spare capital to buy property outright. However, the rental market is still there, as many younger professionals prefer to rent homes because they either cannot afford to buy, or they wish to have the increased flexibility to be able to relocate at relatively short notice. Builders and developers are looking seriously at this option of building homes to rent them out themselves rather than selling to individual investors and landlords.

More work for lawyers

When money is hard to come by, most people become more litigious in an effort to ensure that they get every penny that they feel is due to them. Recessions make contractors become even more aggressive, renegotiations of contracts have increased (possibly as a result of clients feeling that they could have obtained a better price) and the 'claims culture' returns with a vengeance. Paradoxically, the use of adjudication has also increased. This was originally intended as a quick-fix dispute resolution procedure in order to reduce the incidence of more expensive and time-consuming arbitration and litigation. However, many clients who would not have pursued an issue to arbitration or litigation at all are now using this procedure, as it takes only a month or so and the costs are modest. Lawyers also undertake a lot of bankruptcy and insolvency work. As the construction industry normally accounts for the majority of corporate insolvencies in the UK, this figure has clearly increased in times of recession.

Return to single-stage tendering

As we saw in the BAA case study in Chapter 10, many clients consider that traditional competitive tendering is the best way to drive down prices and get better 'value'. The large gaps in many contractors' order books also make them behave more competitively in the estimating and tendering of projects. Clearly, clients are happy to take advantage of this (until they see the claims consultants on site during the construction stage). When the contractors had the advantage, i.e. when demand exceeded supply, then two-stage tendering was the preferred choice. Now the client wants cost certainty and a lump sum before signing up. Desperate contractors cannot really argue in recessionary times, when supply now exceeds demand.

More efficient sites

When contractors are working on a lump-sum fixed price, the only way to increase the profit margin is to reduce costs, and reducing waste is the easiest form of reducing total costs. There are many ways this can be done – keeping the site tidy and materials storage in good order is obviously a primary waste-reduction strategy and most professional contractors are already doing this, but other forms of waste reduction are less easily identifiable or more expensive to implement. More off-site manufacturing and just-in-time materials purchasing can make a significant contribution to physical waste reduction as well as having positive cash-flow benefits.

More cost cutting and making-do.

Recessions always mean hard bargaining over costs, if the project is to go ahead at all. Ideally this would mean value engineering through negotiation, but often it means cutting out a major portion of the work, e.g. the west wing or the top three floors. It may also mean more use of standard products and technologies rather than bespoke products and the use of lower-quality products (i.e. cheaper), with the obvious effect on the product's life and therefore the building's subsequent maintenance costs.

The recession does not just affect construction; client organisations are also affected in their own industries and are therefore understandably more inclined to stay put and adapt their existing premises rather than build new ones. Clearly, if client organisations themselves are being hit by the downturn, their need for space and property will be reduced proportionately, which means more, smaller-scale, remodelling and renovation work but less larger new work. Unoccupied buildings may also be adapted for new uses.

13.2 What are clients now looking for?

In January 2009, the Construction Clients' Group organised a workshop called 'Equal Partners' in order to improve the relationship between customers of the construction industry and the supply chain. The client feedback was that of course the recession has had a major effect on project delays and cancellations and financing has become much more difficult, but it has given an opportunity for clients to differentiate themselves and brought them closer to their preferred supply-chain partners by developing a deeper mutual understanding, a synergy or 'commercial alignment', as the workshop report puts it.

Experienced clients still consider that the industry is too reactive and also that early engagement of the project skills is important. This is linked with the hope that cost reduction should be made by process improvement and innovation, rather than merely cheapening an existing process.

Clients themselves are not without blame, as there are many inexperienced and 'bad' clients out there who have poor decision-making skills and insufficient knowledge of the industry and project processes, as we mentioned in Chapter 2. Clients must also know how to manage the risks – merely passing on all the risks to the contractor will not in itself reduce overall costs.

13.3 Procurement routes best suited to the 'new reality'

13.3.1 Traditional single-stage tendering

As a consequence of the recession, many clients are looking to maximise the element of competition in their tenders by adopting a single-stage competitive strategy. Clients are considering single-stage tendering for the following reasons, many of which are discussed in some length in Chapter 6:

- *The need for greater cost certainty during design and construction.* Some clients who have had problems agreeing a cost plan with the contractor on a multi-stage negotiated contract have also had difficulties controlling costs in the second stage of the tender.
- *The need for a well-documented, fixed-price contract.* The ability to relate a single set of client-side-produced tender documents to the contractor's commercial offer is important for increasingly risk-averse funders.
- *To complete the design before the contractor appointed.* This will allow a fixed price to be calculated with only minor variations and changes during the construction stage.
- *To use commercial pressure to secure cost reductions for projects that might otherwise be unviable.* Changing market conditions have created the opportunity for clients to tender much more competitively. However, most contractors have a range of project work secured on the basis of a mix of procurement routes, and this shift in emphasis does not mean that all work will be won through lowest-price competition. In fact, some contractors and most subcontractors actually prefer competitive bidding, as that is the 'devil they know'.

The willingness of clients to move away from two-stage tendering indicates a degree of frustration with certain aspects of collaborative working. Although some clients may have sufficient workload to be able to engage in collaborations with individual contractors, others have found that two-stage tendering is characterised by tough negotiations in the later stages of the agreement of the price. A harsh economic climate will undoubtedly further encourage adversarial behaviour, but the client side of the industry should recognise that they moved away from single-stage tendering for good reasons. The process can be wasteful of resources, it separates design and construction and, when tendered on incomplete information (which was the case most of the time), provides an illusion of competitive pricing and cost certainty. Just looking at the amount of provisional sums and possibly provisional quantities in a 'firm' BOQ will give a good indication of the cost certainty of the contract sum.

It is therefore essential that clients who adopt the single-stage route should do so in full knowledge of these issues, and with a team capable of completing a proper a design to the level of detail required under the particular form of contract.

13.3.2 Develop and construct

This approach may not be the first choice during a recessionary period as it demands changes in practice from everyone involved, so the most important factors in determining a successful outcome are client buy-in, clear communication and a good contractor-procurement strategy, setting out milestones, reviews, sign-off points, etc. The approach has been successfully adopted on a number of projects, some of which did not have professional construction clients, which illustrates that its use is not restricted to experienced clients, but the clients need to be innovative and reasonably willing to take risks, both of which are in relatively short supply in economic downturns.

By novating the design team to the contractor at a relatively early stage, the client is effectively exchanging control over design development to secure the wider benefits associated with collaborative design and construction. The preliminary design should therefore be developed and the business case signed off before the contractor is appointed and there should also be real engagement with the team from clients and other stakeholders in the later stages of design development, which will be led by the novated contractor. Other success factors include:

- Appointing the right combination of design team, principal contractor and specialist contractors. In addition to their technical skills, their ability to work together as a team will be paramount to the successful outcome of the project as a whole.
- Maintaining well-managed and effective client relations with all members of the team.
- Achieving a complete, co-ordinated design, based on specialist contractor input, at financial close.
- Agreement of an achievable and agreed contract sum with as few provisional sums as possible, therefore allowing the contractors to price for the appropriate risk allowances.
- A total commitment to a collaborative approach and value management principles from all the team.
- A clear procurement strategy setting out roles, dates and procedures and ensuring the client and end-user's engagement in the process.
- Ensuring effective administration of all stages by the project manager, including configuration and change management as well as efficient contract administration.

13.3.3 Construction management

As we saw in Chapter 7, section 7.2, construction management is perhaps best summed up as the procurement route whereby designers design, package contractors detail and deliver the works and where the construction manager provides the site establishment and generally manages the process, leaving the client to lead and accept the risk on a project for which they are ultimately responsible. The key issues affected by the changes to the industry that this definition raises include:

- The central role of the construction manager in managing the project and in providing administrative support to the client.

- There is no single point of responsibility related to the delivery of the project.
- The greater role of trade contractors in the completion of their design work and co-ordination of their work with other packages leads to a further lack of control by the client/main contractor.
- The direct relationship between the client, consultants and trade contractors leads to a fragmentation of communication and control.

Construction management is distinguished by the influence of the client's and construction manager's management and leadership skills on the outcomes of the project, and the potential exposure of the client to loss should these skills prove inadequate. Clearly, problems related to co-ordination and timely decision-making are common to all construction projects, and with construction management especially, it is the performance of the team that contributes most to project success, rather than the procurement strategy itself. This is an area where many clients are now reluctant to take risks, given their very limited financial flexibility.

However, by providing the client with greater influence over the project and more flexibility with contractor selection, construction management also means that lack of performance by the client, because of indecision or delayed instructions, can have a major effect on the project. Furthermore, the consequences of poor management by the construction manager may also directly affect the client, if there were claims for delay and disruption by the trade contractors.

Therefore, the client must be sufficiently disciplined and resourced to provide their input into the project as required by any information-release schedules. They must also ensure that they appoint a construction manager with the appropriate resources and experience (as required by the CDM Regulations anyway) as well as a proactive and positive approach – a task that is challenging for many clients, even in better economic times.

13.4 Building information modelling (BIM)

Building information modelling (BIM) is a relatively new approach to building design, construction and operation and is intended to change the way industry professionals think about how technology can be applied to building design, construction and operation. The intention is that there should be immediate availability and access to project design scope, schedule and cost information that is accurate, reliable, integrated and fully co-ordinated.

Its advantages are seen to be:

- increased speed of delivery of the project (time saved);
- better co-ordination of design (fewer errors);
- decreased costs (money saved);
- increased productivity;
- higher-quality work (right first time);
- better client satisfaction (resulting in new revenue and business opportunities).

All of which relate to our old friends, cost, time and quality – see Chapter 1, section 1.4.

For each of the three major phases in the building life cycle – design, construction and operation – building information modelling offers all the project participants access to the following critical information:

- in the design phase – design, schedule and budget information;
- in the construction phase – quality, schedule and cost information;
- in the operation phase – performance, utilisation and financial information.

Keeping this information up to date and accessible to all who require it in an integrated computerised format is intended to give architects, engineers, contractors and clients a clear overall vision of their projects, as well as the ability to make better-quality decisions, thereby raising the overall quality of the projects, increasing the profitability and improving client satisfaction.

Although building information modelling is essentially a management approach rather than an off-the-shelf technology, it does require suitable technology to be implemented effectively.

Therefore BIM is, essentially, the intersection of two critical ideas:

- Keeping critical design information in digital form in order for it to be easier to update and share and of more value to the organisations creating and using it.
- Creating real-time, consistent relationships between digital design data and the information needed for efficient and effective operation of the building. This can save significant cost in the operation stage.

13.4.1 BIM benefits in the design phase

As we have seen, during the course of a building project, the design team must balance the project scope, schedule and cost. Untimely changes to any of these variables can be expensive and will undoubtedly affect the relationships between the parties to the project. Using traditional methods in the design stage, access to design-related information is usually available directly from the CAD technology, but cost and scheduling information often has to be calculated from this design information using other sources of data, such as pricing information or method statements and programmes. Using BIM techniques, all of this critical information is available in one source, so that project-related decisions can be made more quickly and effectively.

BIM allows a project team to make changes to the project at any time during the design stage without the need for laborious, manual procedures. This gives the team more time to work on design and other high-value design issues. In addition, all of the building design and documentation work can be carried out concurrently instead of consecutively, as design decisions are captured at the point of creation and embedded in the project documentation as the work proceeds.

Whenever a change is considered to the design of a project, all of the consequences of that change are automatically co-ordinated throughout the project. This aspect of configuration management therefore allows the design team to deliver their work faster, because it means that the creation of key project deliverables – such as visualisations and regulatory approval documents – are more automatic and require less time and effort. This automatic co-ordination of changes offered by BIM should eliminate co-ordination mistakes and improve the overall quality of design work, both of which hopefully mean that design companies win more repeat business from satisfied clients.

13.4.2 *BIM benefits in the construction phase*

In the construction phase of the building life cycle, BIM makes available concurrent information on building quality, schedule and cost. The PQS and contractor can accelerate the quantification of the building for estimating and value-engineering purposes and for the production of updated estimates and construction planning. The effects of proposed variations can be studied and understood easily, and the contractor can quickly prepare plans showing site use or phasing for the client, thereby communicating and minimising the impact of construction operations on the client's operations and personnel. BIM also means that less time and money are spent on process and administration issues in construction because document quality is higher and construction planning better. The end result is that more of the owner's hard-earned funds go into the building than into administrative and overhead costs.

13.4.3 *BIM benefits in the operations phase*

In the operations phase of the building life cycle, i.e. when the construction has been completed and the building is 'working', BIM makes available concurrent information on the use or performance of the building; its occupants and contents; the life of the building over time; and the financial aspects of the building, such as rent, taxes and utilities costs. BIM is designed to provide a digital record of renovations and other physical changes to the building, which will help the sale or rental process by providing the documentation required for potential tenants. BIM may also be used to provide data for costs analyses of new buildings where the new buildings are of similar construction and use but in different locations. Physical information about the building, such as finishes, tenant or department assignments, furniture and equipment inventory and financially important data about leasable areas and rental income or departmental cost allocations are all more easily managed and available. Consistent access to these types of information clearly improves both revenue and cost management in the operation of the building.

13.4.4 *Future potential*

When BIM is used effectively, designers will be able to make use of a project's digital design data to provide new services and therefore hopefully gain new sources of income. Experienced clients are increasingly demanding digital models of their buildings, especially where they have a portfolio of buildings in their ownership. Designers and project managers may be able to offer new and expanded services, such as move management, energy analysis, digitally integrated cost estimating and renovation phase planning, no doubt for additional fees.

In the future, BIM is expected to empower both design and construction professionals to work more collaboratively throughout the project delivery process, focusing their energy on more value-added functions such as client requirements, creativity and problem solving, while computers do the tedious tasks of number crunching. Many quantity surveyors will, however, have heard that one before.

But for real property owners and professional project managers, BIM holds great promise beyond merely improving productivity in the design and construction process. Ultimately, this approach and technology has the potential to enable the seamless transfer of knowledge from asset planning through design, construction, facilities management and operation, into the various disposal options. While all parties involved in design and construction stand to gain from the adoption of BIM, it is the clients who will potentially benefit the most, through the use of the facilities model and its embedded knowledge throughout the economic life of the building.

This potential can only be realised if the information contained in the model remains accessible and usable across the variety of technology platforms likely over a long period of time. Given the accelerating pace of technology, in twenty to thirty years our now state-of-the-art hardware and software applications will undoubtedly be outdated and obsolete. It is therefore essential that BIM is developed within a universal, open data standard to allow full and free transfer of data among the various applications.

13.5 Fully integrated project delivery

According to the American Institute of Architects' guide to the delivery of integrated projects, integrated project delivery (IPD) is

> *a project delivery approach that integrates people, systems, business structures and practices into a process that collaboratively harnesses the talents and insights of all participants to optimize project results, increase value to the owner, reduce waste, and maximize efficiency through all phases of design, fabrication, and construction. (AIA 2010)*

IPD principles are designed to be applied to a variety of procurement routes and contractual arrangements, with the IPD teams including members well beyond the basic participants of client, design team and contractor. Integrated projects should expect to have highly effective collaboration between the client, the lead designer and the main constructor, commencing at early design and continuing through to project handover.

According to the above American Institute of Architects' guide, IPD is built on collaboration and trust, as without trust-based collaboration, IPD will falter and participants will remain in the adversarial relationships that 'plague the industry today'. IPD promises better outcomes provided that all project participants embrace the following principles:

- mutual respect and trust
- mutual benefit and reward
- collaborative innovation and decision-making
- early involvement of key participants
- early goal definition
- intensified planning
- open communication
- appropriate technology
- organisation and leadership.

Which all seems remarkably similar to the discussion on partnering in Chapter 9. The major difference here is that all stakeholders are involved in IPD, not just the construction stage participants, but the facilities managers, tenants, etc. In addition, the use of appropriate technologies, such as BIM, is expected.

13.6 Virtual design and construction (VDC)

VDC is essentially an extension of computer-aided design (CAD), in that the digital design is fed into project management software and business process software, to assess the best method and sequence of construction and to additionally assess realistic times and costs of construction. This may be a very simplified analysis and will not be extended here; further detail in this area can be found in various technical academic papers and web-based articles available on the art and science of VDC. For present purposes it is enough to know that, according to the Stanford University Working Paper on VDC (Kunz and Fischer 2009), the basis of VDC includes several major components, which include:

- *Engineering modelling methods*, which represent the product, organisation, and process.
- *Model-based analysis methods*, which predict the project schedule, cost, effort, hidden work, organisation, process and schedule risks. These methods require a CAD input in 3D, which is then converted to a 4D model (which apparently means the 3D model being built over time, time being the fourth dimension).
- *Visualisation methods*, to present views of the product, organisation and process in ways that are clear for normal human construction professionals as well as a broad class of interested stakeholders.
- *Business metrics and methods*, to manage project processes using measured performance indicators.
- *Economic impact*; that is quantitative models of both cost and value of capital investments, including the project as a whole, individual project elements and any incremental investments required to change the process.

13.7 Sustainability and environmental issues

The economic downturn could not have come at a worse time for the environment. Just as government and developers had spent a fortune on their environmental policies, the credit crunch has made everyone pause for breath and concentrate their energies on the more critical element of survival. For most developers, the 'green' elements of the projects have often been the first to go. Environmentally friendly designs are a good augmentation to a project, but they are rarely central to a purchaser's or tenant's decision processes. Despite this, the emergence of sustainable development concepts could hold the key for both producers and consumers of building products and services. Engaging with the three pillars of sustainable development will involve facets that are environmental, economic and social – and may enable long-term survival for all stakeholders in the industry.

13.7.1 What is sustainable construction?

The UK government produced a strategy for sustainable construction in June 2008, helpfully entitled 'Strategy for Sustainable Construction' (BERR 2008). Sections of the report are concerned with the means of achieving sustainable construction:

- *Procurement* – encouraging better supply-side integration and commitments.
- *Design* – ensuring that projects are buildable, fit for purpose, resource efficient, sustainable, resilient, adaptable and attractive.
- *Innovation* – to increase the sustainability of both the construction process and its resultant assets.
- *People* – continuous improvement in skills training, Investors in People, CPD and Life Long Learning.
- *Better regulation* – a reduction in the administrative burden affecting public, private and not-for-profit developments.

The report also deals with the 'end product' which will define sustainable construction:

- *Climate change mitigation* – reduction in total CO_2 emissions of both the construction process and also the building in use.
- *Climate change adaptation* – change the procedures so that more carbon neutral policies are the norm.
- *Water* – development of new technologies which reduce per capita water consumption, by, for example, more water reuse ('grey' water for gardens) or water-use reduction (short-flush toilets).
- *Biodiversity* – conservation and biodiversity around construction and development sites.
- *Waste* – a major reduction in construction related waste to landfill sites.
- *Materials* – construction materials to be more environmentally friendly with less embodied energy.

13.8 Summary and tutorial questions

13.8.1 Summary

The longer-term effects of the economic downturn on the construction industry

- Since 2008, the commercial property sector has had a steep fall and is unlikely to recover for several more years.
- The initial principal cause of the recession was a crisis in the banking industry largely caused by injudicious lending.
- Credit restriction and low liquidity for business has been a result of the fallout.
- Government is likely to target public-sector construction in their austerity programme, so any public-sector-driven construction work is not necessarily safe.

- Small businesses in the building industry do not have larger projects to keep them afloat during difficult times.
- Procurement has seen a swing back towards single-stage, lowest-bid tenders aimed at achieving the lowest possible out-turn cost.
- Builders and developers are looking seriously at build-to-let instead of buy-to-let homes by renting them out themselves rather than selling to individual investors and landlords.
- The construction industry normally accounts for the majority of corporate insolvencies in the UK. This figure and litigation in general have clearly increased in times of recession.
- There is a return to single-stage tendering, more efficient sites and more cost cutting.

What are clients now looking for?

- The downturn has given an opportunity for clients to differentiate themselves and brought them closer to their preferred supply-chain partners.
- Clients must know how to manage the risks – merely passing on all the risks to the contractor will not in itself reduce overall costs.

Procurement routes best suited to the 'new reality'

- Many clients are looking to maximise the element of competition in their tenders by adopting a single-stage competitive strategy.
- The willingness of clients to move away from two-stage tendering indicates a degree of frustration with certain aspects of collaborative working.
- The most important factors in determining a successful outcome are client buy-in, clear communication and a good contractor-procurement strategy.
- By novating the design team to the contractor at a relatively early stage, the client is effectively exchanging control over design development to secure the wider benefits associated with collaborative design and construction.
- It is important to appoint a construction manager with the appropriate resources and experience as well as a proactive and positive approach – a task that is challenging for many clients, even in better economic times.

Building information modelling (BIM)

- BIM is a relatively new approach to building design, construction and operation and is intended to change the way industry professionals think about how technology can be applied to building design, construction, and operation.
- There should be immediate availability and access to project design scope, schedule and cost information that is accurate, reliable, integrated and fully co-ordinated.
- BIM benefits the design, operation and construction phases of a building project.
- Information technology has the potential to enable the seamless transfer of knowledge from asset planning through design, construction, facilities management and operation, into the various disposal options.
- It is essential that BIM is developed within a universal, open data standard to allow full and free transfer of data among the various applications.

Fully integrated project delivery

- IPD principles are designed in application to a variety of procurement routes and contractual arrangements.
- Integrated projects should expect to have highly effective collaboration between the client, the lead designer and the main constructor, commencing at early design and continuing through to project handover.

Virtual design and construction (VDC)

- VDC is essentially an extension of computer-aided design (CAD).
- In VDC the digital design is fed into project management software and business process software, to assess the best method and sequence of construction and to additionally assess realistic times and costs of construction.
- VDC components include methods in engineering modelling, model-based analysis, visualisation, business metrics and economic impact.

Sustainability and environmental issues

- Environmentally sound designs are a good augmentation to a project, but they are rarely central to a purchaser's or tenant's decision processes.
- Engaging with the three pillars of sustainable development will involve facets that are environmental, economic and social – and may enable long-term survival for all stakeholders in the industry.
- The UK government Strategy for Sustainable Construction (2008) aims to achieve sustainable construction in procurement, design, innovation, people, and better regulation.
- The end products that will define sustainable construction are those responding to climate change mitigation and adaptation, water, biodiversity, waste and materials.

13.8.2 Tutorial questions (essays)

1 How has the recent economic downturn and credit crisis affected building industry procurement practices?
2 How can procurement play a part in the survival of the building industry over the next decade?
3 Define building information management, fully integrated project delivery and virtual design and construction. Discuss which is the most appropriate procurement approach in contemporary building projects.
4 Is sustainable construction a utopian or realistic vision? Discuss using examples of procurement in the building industry.

References and further reading

Akintoye, A. and Beck, M. (eds). (2009). *Policy, Finance and Management for Public–Private Partnerships*. Oxford: Blackwell.

Allen, G. (2001). *The Private Finance Initiative (PFI)*. House of Commons Library Research Paper 01/117. Economic Policy and Statistics Section, House of Commons Library.

American Institute of Architects (AIA). (2010). *Integrated Project Delivery: A Guide*. Washington, DC: AIA.

Arthur Anderson and LSE Enterprise. (2000). *Value for Money Drivers in the Private Finance Initiative*. Arthur Anderson and LSE Enterprise.

Banwell, H. (1964). *Report of the Committee on the Placing and Management of Contracts for Building and Civil Engineering Work*. HMSO.

Bennett, J. and Flanagan, R. (1983). 'For the good of the client', *Building*, 27, pp. 26–7.

Bennett, J. and Jayes, S. (1998). *Seven Pillars of Partnering: A Guide to Second Generation Partnering*. Reading Construction Forum.

Blackwell, M. (2000). *The PFI/PPP and Property*. Oxford: Chandos Publishing.

British Property Federation (BPF). (1983). *System for Building Design and Construction*. BPF.

Building. (2006). 'Procurement: public sector projects', *Building*, issue 47.

Building. (2009). 'It's back!', *Building*, 6 November.

Cabinet Office Efficiency Unit. (1995). *Construction Procurement by Government*. HMSO.

Chevin, D. (2002). 'Only themselves to blame', *Building*, 1 February.

Clough, R., Sears, G. and Sears, S.K. (2000). *Building Project Management*. New York: Wiley.

Construction Industry Board (CIB). (1997). *Partnering in the Team*. Thomas Telford Publishing.

Cristobal, J. (2009). 'Time, cost, and quality in a road building project', *Journal of Construction Engineering and Management*, 135 (11), pp. 1271–4.

Cyert, R.M. and March, J.G. (1963). *A Behavioral Theory of the Firm*. Englewood Cliffs, NJ: Prentice-Hall.

Department for Business, Enterprise & Regulatory Reform (BERR). (2008). *Strategy for Sustainable Construction*. HMSO.

Department of the Environment, Transport and the Regions (DETR). (2000). *KPI Report for The Minister for Construction*. The KPI Working Group and HMSO.

Department of Trade and Industry. (1998). *Rethinking Building: Report of the Building Task Force* [Egan Report]. HMSO.

Doran, H. (2001). *Sector Analysis: Building*. English Nature.

Economic Development Committee for Building (EDCB). (1967). *Action on the Banwell Report*. HMSO.

Eggleston, B. (2006). *The NEC3 Engineering and Construction Contract: A Commentary*, 2nd edn. Oxford: Blackwell Science.

Emmerson, H.C. (1962). *Survey of Problems before the Construction Industries*. HMSO.

European Commission. (2010). *Initiative on Public Private Partnerships and Community Law on Public Procurement and Concessions*. Luxembourg: Office of the EC.

Fox, J. and Tott, N. (1999). *The PFI Handbook*. Bristol: Jordan Publishing.

Franks, J. (1990). *Building Procurement Systems – A Guide to Building Project Management*. Ascot: Chartered Institute of Building.

Higgin, G. and Jessop, N. (1965). *Communications in the Building Industry: The Report of a Pilot Study*. Tavistock Publications.

HM Treasury. (2000). *Public Private Partnerships: The Government's Approach*. HMSO.

HM Treasury Central Unit on Purchasing. (1992). *Guidance no. 36 Contract Strategy: Selection for Major Projects*. HMSO.

HM Treasury Taskforce. (2000). *Step by Step Guide to the PFI Process*. Private Finance Unit, HM Treasury.

HMSO. (1950). *The Working Party Report to the Minister of Works: The Phillips Report on Building (1948–50)*. HMSO.

HMSO. (1974). *Health and Safety at Work, etc. Act 1974*. HMSO.

HMSO. (1992). *Management of Health and Safety at Work Regulations 1992*. HMSO.

HMSO. (1994a). *Constructing the Team. Final Report of the Government/Industry Review of Procurement and Contractual Arrangements in the UK Construction Industry* [Latham Report]. HMSO.

HMSO. (1994b). *The Construction (Design and Management) Regulations 1994*. HMSO.

HMSO. (1996a). *Government Contract (GC)/Works/1: Edition Three, Single Stage Design and Build Version*. HMSO.

HMSO. (1996b). *Housing Grants, Construction and Regeneration Act 1996*. HMSO.

IFSL Research. (2009). *PFI in the UK and PPP in Europe 2009*. Sage.

Institution of Civil Engineers (ICE). (1992). *Design and Construct Form of Contract*. ICE.

International Federation of Consulting Engineers (FIDIC). (2010). *Standard Form for EPC/Turnkey Projects: The 'Silver Book'*. Geneva: FIDIC.

Joint Contracts Tribunal (JCT). (2002). *Practice Note 6 Main Contract Tendering*. JCT Ltd.

Joint Contracts Tribunal (JCT). (2005a). *Standard Form of Design Build Contract (DB)*. JCT Ltd.

Joint Contracts Tribunal (JCT). (2005b). *JCT05 Framework Agreement (FA)*. JCT Ltd.

Joint Contracts Tribunal (JCT). (2008) *Practice Note: Deciding on the Appropriate JCT Contract*. JCT Ltd.

Joint Contracts Tribunal (JCT). (2009). *Standard Building Contract with Approximate Quantities (SBC/AQ) Revision 2*. JCT Ltd.

Jones, P. and Evans, J. (2008). *Urban Regeneration in the UK*. Sage.

Kunz, J. and Fischer, M. (2009) *Virtual Design and Construction: Themes, Case Studies and Implementation Suggestions*. Stanford University CIFE Working Paper 97, October. Stanford, CA: Centre for Integrated Facility Engineering, Stanford University.

Leibenstein, H. (1973). 'Competition and *x*-inefficiency: reply', *Journal of Political Economy*, 81, pp. 765–77.

Masterman, J.W.E. (2002). *Introduction to Building Procurement Systems*, 2nd en. Spon Press.

Matthews, J., Tyler, A. and Thorpe, A. (1996). 'Pre-construction project partnering: developing the process', *Engineering Construction and Architectural Management*, 3 (1/2), pp. 117–31.

McGraw-Hill Construction. (2010). *Network Help: Action Stages*. New York: McGraw-Hill Construction.

Ministry of Works. (1944). *Report of the Management and Planning of Contracts* [Simon Report]. HMSO.

Mumford, M. (1998). *Public Projects, Private Finance: Understanding the Principles of the Private Finance Initiative*. Griffin.

Myers, D. (2008). *Building Economics: A New Approach*. Taylor & Francis.

Naoum, S.F. and Langford, D. (1987). 'Managing contractors – the client's view', *Journal of Construction Engineering and Management*, 113 (3), pp. 368–84.

National Audit Office (NAO). (2001). *Managing the Relationship to Secure a Successful Partnership in PFI Projects*. The Stationery Office.

National Audit Office (NAO). (2010). *Value for Money Reports on PPP/PFI*. NAO.

National Economic Development Office (NEDO). (1975). *The Public Client and the Construction Industry* [Wood Report]. HMSO.

National Economic Development Office (NEDO). (1978). *Construction for Industrial Recovery*. HMSO.

National Economic Development Office (NEDO). (1988). *Faster Building for Industry*. HMSO.

National Economic Development Office (NEDO). (1991). *Partnering: Contracting without Conflict*. HMSO.

National Joint Consultative Committee for Building (NJCC). (1995). *Code of Procedure for Selective Tendering for Design and Build*. NJCC.

Office for National Statistics (ONS). (2006). *Construction Industry Key Performance Indicators*. HMSO.

Office for National Statistics (ONS). (2009). *Building Statistics Annual: Output*. ONS.

Office for National Statistics (ONS). (2010). *Construction Statistics Annual 2010*. ONS.

Office of Government Commerce (OGC). (2010). *European Procurement Directive*. OGC.

Public Private Finance. (2009a). 'PFIs "Worst year in a decade"', *Public Private Finance*, 130, p. 7.

Public Private Finance. (2009b). 'Funding worries and more delays shadow Building Schools scheme', *Public Private Finance*, 130, p. 6.

Public Private Finance. (2009c). 'The big hiccup', *Public Private Finance*, 130, p. 3.

Rawlinson, S. and Langdon, D. (2010a). 'Procurement: building services', *Building*, 11 May 2007.

Rawlinson, S. and Langdon, D. (2010b). 'Procurement: competitive dialogue', *Building*, 23 May 2008.

Reading Construction Forum (RCF). (1995). *Trusting the Team: The Best Practice. Guide to Partnering in Construction*. Reading Construction Forum.

Reading Construction Forum (RCF). (1998). *The Seven Pillars of Partnering*. Reading Construction Forum.

Royal Institute of British Architects (RIBA). (2007). *The RIBA Outline Plan of Work*. RIBA.

Royal Institution of Chartered Surveyors (RICS). (1987). *JCT Management Contract*. RICS.

Royal Institution of Chartered Surveyors (RICS). (1988). *SMM7: The Standard Method of Measurement of Building Work – Seventh Edition*. RICS.

Royal Institution of Chartered Surveyors (RICS). (1991). *CESMM3: The Civil Engineering Standard Method of Measurement – Third Edition*. RICS.

Royal Institution of Chartered Surveyors (RICS). (2005). *NEC3 Engineering and Construction Contract Option F Management Contract*. RICS.

Royal Institution of Chartered Surveyors (RICS). (2010a). *Contracts in Use Survey*. RICS.

Royal Institution of Chartered Surveyors (RICS). (2010b). *New Rules of Measurement (NRM)*. RICS.

Slough Estates. (1976). *Industrial Investment. A Case Study in Factory Building*. Slough Estates.

Slough Estates. (1979). *Industrial Investment. A Case Study in Factory Building*. Slough Estates.

Trowers and Hamlins. (2005). *PPC2000: The ACA Standard Form of Contract for Project Partnering: Briefing Note*. Trowers and Hamlins.

University of Reading, Department of Construction Management. (1979). *UK and US Construction Industries: A Comparison of Design and Contract Procedures*. Royal Institution of Chartered Surveyors.

Wall, A. and Connolly, C. (2009). 'The Private Finance Initiative: an evolving agenda', *Public Management Review*, 11 (5), pp. 704–24.

WorkSafe BC. (2010). *Part 20 Building, Excavation and Demolition*. Vancouver: WorkSafe BC.

Useful websites

Building magazine: http://www.building.co.uk/.
Constructing Excellence: http://www.constructingexcellence.org.uk/.
Construction News: http://www.cnplus.co.uk/.
Contract Journal: http://www.contractjournal.com/.
Federation of Property Societies (FPS): http://www.fedps.org.uk/.
National Health Service (NHS). NHS Procure21 Framework Agreement (2010): http://www.nhs-procure21.gov.uk/.
Office of Fair Trading (OFT): http://www.oft.gov.uk/.
Office of Government Commerce (OGC): http://www.ogc.gov.uk/index.asp.
Official Journal of the European Union (*OJEU*): http://www.ojec.com/.

Index